Organ Printing

Organ Printing

**Dong-Woo Cho, Jung-Seob Lee, Jinah Jang,
Jin Woo Jung and Jeong Hun Park**
*Department of Mechanical Engineering,
Pohang University of Science and Technology (POSTECH),
77 Cheongam-Ro, Nam-gu, Pohang, Kyungbuk 790-784, Korea*

Falguni Pati
*Department of Biomedical Engineering,
Indian Institute of Technology Hyderabad, Kandi, Sangareddy - 502285,
Telangana, India*

Morgan & Claypool Publishers

Rights & Permissions
To obtain permission to re-use copyrighted material from Morgan & Claypool Publishers, please contact info@morganclaypool.com.

ISBN   978-1-6817-4079-9 (ebook)
ISBN   978-1-6817-4015-7 (print)
ISBN   978-1-6817-4207-6 (mobi)

DOI   10.1088/978-1-6817-4079-9

Version: 20151001

IOP Concise Physics
ISSN 2053-2571 (online)
ISSN 2054-7307 (print)

A Morgan & Claypool publication as part of IOP Concise Physics
Published by Morgan & Claypool Publishers, 40 Oak Drive, San Rafael, CA, 94903, USA

IOP Publishing, Temple Circus, Temple Way, Bristol BS1 6HG, UK

Dedicated to the tissue/organ regeneration community.

Contents

Preface

Organ printing offers exciting prospects for fabricating complex three-dimensional (3D) multicellular organs *in vitro* by integrating technologies from the fields of engineering, biomaterials science, cell biology, physics, and medicine. The well-organized tissue/organ constructs can be printed via 3D printing technology by delivering precisely characterized progenitor cell populations and suitable biomaterials in a defined and organized manner, at a targeted location, in adequate numbers, and within the right environment for repairing or replacing defective human tissues and organs.

3D printing technology enables the fabrication of desired complex 3D shapes through a layer-by-layer deposition process and mimicry of the native tissue microenvironment with precisely controlled positioning of ultra-low volumes of biomaterials along with cells on a micro- and/or nano-scale. Various 3D printing systems for organ printing have been developed that can be classified into three major classes: stereolithography-based, inkjet-based, and dispensing-based 3D printing. All these 3D printing methods use computer-aided design/computer-aided manufacturing (CAD/CAM) technology to realize the desired 3D shape of tissue/organ constructs.

Cells and biomolecules are usually printed by encapsulating them in various dispensable hydrogels, and the mixture is typically called 'bioink'. The role of bioink is to protect the cells from harsh environments during the printing process, as well as to enhance cellular functions, such as proliferation, differentiation, and maturation. After printing a 3D structure using bioink, it is then stabilized by thermal, ionic, or chemical crosslinking methods. There are several hydrogels that are commonly used in bioinks and they can largely be classified into three kinds: natural-derived, synthetic, and semi-synthetic materials. Decellularized extracellular matrix-derived materials are a promising source of natural-derived bioinks that provide excellent biocompatibility and biofunctionality.

Organ printing is being applied in the tissue engineering field with the purpose of developing tissue/organ constructs for the regeneration of both hard (bone, cartilage, osteochondral) and soft tissues (heart). There are other potential application areas including tissue/organ models, disease/cancer models, and models for physiology and pathology, where *in vitro* 3D multicellular structures developed by organ printing are valuable.

This book introduces various 3D printing systems, biomaterials, and cells for organ printing. In view of the latest applications of several 3D printing systems, their advantages and disadvantages are also discussed. A basic understanding of the entire spectrum of organ printing provides pragmatic insight into the mechanisms, methods, and applications of this discipline.

Dong-Woo Cho*, POSTECH, Korea
September 2015

* Corresponding author, e-mail: dwcho@postech.ac.kr

Acknowledgements

The authors would like to acknowledge the support and dedication of the many colleagues and friends in the field of biofabrication, tissue engineering and regenerative medicine. This work was supported by the National Research Foundation of Korea (NRF) grant funded by the Korea government (MSIP) (No 2010-0018294). Finally, the authors appreciate the help and support extended by members of the IMS Lab at POSTECH.

Author biography

Dong-Woo Cho, Jung-Seob Lee, Falguni Pati, Jin Woo Jung, Jinah Jang and Jeong Hun Park

Professor Dong-Woo Cho

Professor Dong-Woo Cho received his PhD in Mechanical Engineering from the University of Wisconsin-Madison in 1986. Since then he has been a professor of the Department of Mechanical Engineering at the Pohang University of Science and Technology. He is director of the Center for Rapid Prototyping-based 3D Tissue/ Organ printing. His research interests include 3D microfabrication based on 3D printing technology, its application to tissue engineering, and more generally to bio-related fabrication. He has recently focused on tissue/organ printing technology and the development of high-performance bioinks. He has received several prestigious awards in these academic areas. He serves or has served on the editorial boards of several international journals. Professor Cho has published over 210 academic papers in various international journals in the field of manufacturing and tissue engineering, and has contributed chapters to ten books related to tissue engineering and organ printing.

Jung-Seob Lee

Jung-Seob Lee is an Integrated MS/PhD student in the Department of Mechanical Engineering at Pohang University of Science and Technology (POSTECH). He received his BS at the Department of Mechanical Engineering at Busan University in 2010. His research interests is the development and improvement of 3D printing systems and its application to the regeneration of large volume composite tissue with complex shapes. His current research is to develop a 3D printed ear and to regenerate auricular cartilage for microtia patients.

Professor Falguni Pati

Falguni Pati is an Assistant Professor in the Department of Biomedical Engineering at IIT Hyderabad, India. He received his ME in Biomedical Engineering from Jadavpur University, India and completed his PhD in Bioengineering from IIT Kharagpur, India. He was a Postdoctoral Fellow at Intelligent Manufacturing Systems Lab, POSTECH, South Korea from 2011 to 2014 and in the Division of Nanobiotechnology at the Royal Institute of Technology (KTH) from 2014 to 2015. His research interests include 3D bioprinting, tissue/organ models, novel bioinks, and tissue engineering. He has published 15 peer reviewed journal articles and four book chapters. He has also applied for two patents with his co-workers.

Jin Woo Jung

Jin Woo Jung is an Integrated MS/PhD student in the Department of Mechanical Engineering at Pohang University of Science and Technology (POSTECH). He received his BS at the Department of Mechanical Engineering, Hanyang University in 2009. His research interests are the development and improvement of 3D printing and its CAD/CAM technology. His current research is in 3D printing-based clinical applications, such as the reconstruction of a new nose for an arhinia patient, reconstruction of depressed malar region, and development of customized augmentation rhinoplasty using 3D printing systems.

Jinah Jang

Jinah Jang is a postdoctoral fellow in the Department of Mechanical Engineering at Pohang University of Science and Technology (POSTECH). She received her BSc(Eng) in Mechanical Design and Automation Engineering at Seoul National University of Technology in 2010 and her PhD in the Division of Integrative Biosciences and Biotechnology at POSTECH in 2015. Her current research interests are to develop the functional tissue-derived decellularized extracellular matrix (dECM) bioinks for 3D bioprinting of tissue analogues. In particular, she applies this technology to both regenerative medicine and *in vitro* tissue modeling for cardiovascular disease. She has received several awards including excellent presentation, excellent publication and young investigator presentation awards in 2014–2015.

Jeong Hun Park

Jeong Hun Park is a postdoctoral fellow in the Department of Mechanical Engineering at Pohang University of Science and Technology (POSTECH), Korea from 2015. He received his BS and ME in Precision Mechanical Engineering at Chungbuk National University, and completed his PhD in Mechanical Engineering at POSTECH. His current research interests are the development of 3D printing systems, 3D printing of target-specific functional scaffolds using various biomaterials, and the building of 3D tissue/organ constructs for tissue regeneration.

IOP Concise Physics

Organ Printing

Dong-Woo Cho, Jung-Seob Lee, Falguni Pati, Jin Woo Jung, Jinah Jang and Jeong Hun Park

Chapter 1

Introduction

Organ printing is essentially based on 3D printing, or additive manufacturing, which is the automated computer-aided deposition of living cells, together with matrix materials and biochemical factors, at specified positions with adequate numbers and in the right combination for the development of three-dimensional (3D) tissue/organ constructs [1, 2]. Charles W Hull first described a 3D printing method in 1986, which he named 'stereolithography' [3]. In this process, thin layers of a material are printed sequentially in layers and subsequently cured with ultraviolet light to form a solid 3D structure. However, this process was used to create sacrificial resin molds, which were then used for making 3D scaffolds from biological materials. Direct printing of biological materials into 3D scaffolds was made possible after the development of solvent-free, aqueous-based systems, and these scaffolds were used for transplantation with or without seeded cells [4]. True 3D cell printing, where living cells are included in the printing process itself, was made possible by more recent advances in 3D organ printing technology, cell biology and materials science [5].

In organ printing, layer-by-layer precise positioning of living cells, matrix materials, biochemicals, and bioactive factors at a deliberately targeted location with a resolution similar to that found in biological tissues is used to fabricate 3D structures [6]. Researchers are trying to print living tissues with biological and mechanical properties suitable for the clinical restoration of tissue and organ function [2]. However, the key challenge is to replicate the intricate micro-architecture of the extracellular matrix (ECM) components and the organization of multiple cell types at a sufficient resolution to develop functional tissues or organs [7].

Despite the initial challenges, organ printing has experienced rapid growth in the last few years, due to the development of novel tools and technologies. Although whole organ development is still at an early stage, this technique has the potential to deliver a long-term clinical solution to the tissue and organ shortage [8]. Implants with anatomical shapes and sizes can be made using computer-aided design (CAD) combined with medical imaging techniques, such as computed tomography (CT)

doi:10.1088/978-1-6817-4079-9ch1   1-1

and magnetic resonance imaging (MRI) [9]. In recent years there has been increasing interest in applying this technology to various applications in biology and medicine. The development of diverse 3D organ printing technologies has also enabled the inclusion of highly sensitive stem cells. This can be seen from the use of various kinds of stem cells, such as human bone marrow stem cells (BMSCs), adipose-derived stem cells (ASCs), and even highly sensitive embryonic stem cells (ESCs) [10–12]. Interestingly, 3D organ printing can be advantageous for controlling microenvironments within a structure through the generation of spatial gradients of immobilized macromolecules to direct the fate of stem cells [13–15].

Based on the working principle, organ printing systems can be primarily classified as stereolithography-based, inkjet-based, and dispensing-based. Inkjet-based organ printing is one of the primary and the most promising biofabrication approaches currently available [16]. Inkjet printing is a non-contact technique that works by depositing ink drops in successive layers, at times with the support of a biopaper, to produce biological tissues or organs [4, 17, 18]. Laser-assisted bioprinting (LaB) uses a pulsed laser source, an absorption layer and a substrate to directly position multiple cells and biological components onto arbitrary surfaces to fabricate living tissues or organs [19]. Although inkjet- and stereolithography-based organ printing technologies offer great potential for printing living cells onto target-specific positions, the printing of clinically relevant 3D tissues or organs is hard to achieve with these technologies [20]. Dispensing-based organ printing technologies, using a syringe, micronozzle and pressure system, seem to be the most promising approach for producing clinically relevant-sized 3D tissue or organ constructs [21].

Several kinds of biomaterials are used for the 3D organ printing of tissue and organ constructs. Biopolymers are widely used biomaterials for the organ printing of scaffolds as well as living tissue/organ constructs. They offer distinct advantages in terms of biocompatibility, versatility of chemistry and the biological properties that are important in tissue engineering and regenerative medicine [22]. Biopolymers can be classified into several categories with respect to their structural, chemical and biological characteristics, and based on their origin (e.g., naturally occurring, synthetic or semi-synthetic) [22]. Synthetic biopolymers such as polycaprolactone (PCL), polylactic acid (PLA) and poly(lactic-co-glycolic acid) (PLGA) are used for printing structurally stable 3D scaffolds for implantation with or without seeded cells [23–25]. To impart light sensitivity, biopolymers of both natural and synthetic origin have been modified to attach chemical groups that induce crosslinking upon exposure to light [26]. Acrylated biopolymers, such as polyethylene (glycol) diacrylate (PEGDA) [27] and gelatin methacrylate [28], have been used for printing 3D constructs using stereolithography-based 3D printers. Hydrogels are popular for encapsulating cells because they provide an aqueous gel environment. Cells encapsulated in hydrogels (often called 'bioink') are generally fed into the dispensing-based 3D printer to make cell-laden constructs [29]. These are solidified through thermal processes or post-print crosslinking and used to produce diverse tissues ranging from liver to bone using materials such as gelatin [30], gelatin/chitosan [31], gelatin/alginate [32], gelatin/fibrinogen [33], Lutrol F127/alginate [34] and alginate [20]. However, these materials do not represent the complexity of natural ECMs and are thus inadequate to recreate tissue-specific microenvironments. Consequently, the cells within these hydrogels do not exhibit the intrinsic morphologies and functions of living tissues *in vivo*. Recently, Cho's group developed bioink from decellularized extracellular matrices and used it to organ print 3D tissue analogs [8]. The decellularized extracellular matrix (dECM) provides a natural chemical milieu for the embedded cells because, to date, no other natural or man-made material has been found to fully recapitulate all the features of natural ECM [35].

There are several application areas where bioprinted tissues/organs are valuable, such as tissue engineering [36, 37], cell-based sensors [38], drug/toxicity screening [39] and tissue and tumor models [40]. An example of the application of 3D organ printing to produce medical devices for use in the clinic is the bioresorbable airway splint [41]. *In vitro* models of human physiology and pathology can also be developed using bioprinted tissues/organs and can convincingly and accurately predict the outcome of *in vivo* drug administration and potential toxic exposure in human-specific models [39]. Furthermore, cancer models can be developed with organ printing, recreating the microenvironmental characteristics representative of tumors *in vivo* [40].

In this book, we describe the principles and processes of several tissue/organ printing methods. We also discuss various synthetic materials and hydrogels used for organ printing, together with their processing conditions. We review the applications of 3D organ printing in tissue and organ engineering and develop *in vitro* models for drug discovery and cancer models. Organ printing's achievements, challenges and future directions are also discussed.

References

[1] Mironov V, Kasyanov V, Drake C and Markwald R R 2008 Organ printing: promises and challenges *Regen. Med.* **3** 93–103

[2] Moon S *et al* 2010 Layer by layer three-dimensional tissue epitaxy by cell-laden hydrogel droplets *Tissue Eng.* **C 16** 157–66

[3] Hull C W 1986 Apparatus for production of three-dimensional objects by stereolithography US 4575330 A

[4] Nakamura M, Iwanaga S, Henmi C, Arai K and Nishiyama Y 2010 Biomatrices and biomaterials for future developments of bioprinting and biofabrication *Biofabrication* **2** 014110

[5] Ferris C, Gilmore K, Wallace G and Panhuis M 2013 Biofabrication: an overview of the approaches used for printing of living cells *Appl. Microbiol. Biotechnol.* **97** 4243–58

[6] Nakamura M *et al* 2005 Biocompatible inkjet printing technique for designed seeding of individual living cells *Tissue Eng.* **11** 1658–66

[7] Murphy S V and Atala A 2014 3D bioprinting of tissues and organs *Nat. Biotech.* **32** 773–85

[8] Pati F *et al* 2014 Printing three dimensional tissue analogues with decellularized extracellular matrix bioink *Nat. Commun.* **5** 3935

[9] Ballyns J J and Bonassar L J 2009 Image-guided tissue engineering *J. Cell. Mol. Med.* **13** 1428–36

[10] Gruene M *et al* 2011 Laser printing of three-dimensional multicellular arrays for studies of cell–cell and cell–environment interactions *Tissue Eng.* **C 17** 973–82

[11] Gruene M *et al* 2011 Laser printing of stem cells for biofabrication of scaffold-free autologous grafts *Tissue Eng.* **C 17** 79–87

[12] Koch L *et al* 2010 Laser printing of skin cells and human stem cells *Tissue Eng.* **C 16** 847–54

[13] Ker E D *et al* 2011 Engineering spatial control of multiple differentiation fates within a stem cell population *Biomaterials* **32** 3413–22

[14] Ker E D *et al* 2011 Bioprinting of growth factors onto aligned sub-micron fibrous scaffolds for simultaneous control of cell differentiation and alignment *Biomaterials* **32** 8097–107

[15] Phillippi J A, Miller E, Weiss L, Huard J, Waggoner A and Campbell P 2008 Microenvironments engineered by inkjet bioprinting spatially direct adult stem cells toward muscle- and bone-like subpopulations *Stem Cells* **26** 127–34

[16] Boland T, Xu T, Damon B and Cui X 2006 Application of inkjet printing to tissue engineering *Biotechnology J.* **1** 910–7

[17] Xu T, Jin J, Gregory C, Hickman J J and Boland T 2005 Inkjet printing of viable mammalian cells *Biomaterials* **26** 93–9

[18] Nishiyama Y *et al* 2009 Development of a three-dimensional bioprinter: construction of cell supporting structures using hydrogel and state-of-the-art inkjet technology *J. Biomech. Eng.* **131** 035001

[19] Catro S *et al* 2011 Laser-assisted bioprinting for creating on-demand patterns of human osteoprogenitor cells and nano-hydroxyapatite *Biofabrication* **3** 025001

[20] Fedorovich N E *et al* 2012 Biofabrication of osteochondral tissue equivalents by printing topologically defined, cell-laden hydrogel scaffolds *Tissue Eng.* C **18** 33–44

[21] Shim J-H, Lee J-S, Kim J Y and Cho D-W 2012 Bioprinting of a mechanically enhanced three-dimensional dual cell-laden construct for osteochondral tissue engineering using a multi-head tissue/organ building system *J. Micromech. Microeng.* **22** 085014

[22] Lee H B, Khang G and Lee J H 2003 Polymeric biomaterials *Biomaterials: Principles and Applications* ed J B Park and J D Bronzino (Boca Raton, FL: CRC)

[23] Kim J Y and Cho D W 2009 Blended PCL/PLGA scaffold fabrication using multi-head deposition system *Microelectron. Eng.* **86** 1447–50

[24] Kim J Y, Park E K, Kim S-Y, Shin J-W and Cho D-W 2008 Fabrication of a SFF-based three-dimensional scaffold using a precision deposition system in tissue engineering *J. Micromech. Microeng.* **18** 055027

[25] Kim M S, Kim J H, Min B H, Chun H J, Han D K and Lee H B 2011 Polymeric scaffolds for regenerative medicine *Polym. Rev.* **51** 23–52

[26] Guillotin B *et al* 2010 Laser assisted bioprinting of engineered tissue with high cell density and microscale organization *Biomaterials* **31** 7250–6

[27] Tan G, Wang Y, Li J and Zhang S 2008 Synthesis and characterization of injectable photocrosslinking poly (ethylene glycol) diacrylate based hydrogels *Polym. Bull.* **61** 91–8

[28] Billiet T, Gevaert E, De Schryver T, Cornelissen M and Dubruel P 2014 The 3D printing of gelatin methacrylamide cell-laden tissue-engineered constructs with high cell viability *Biomaterials* **35** 49–62

[29] Melchels F P W, Domingos M A N, Klein T J, Malda J, Bartolo P J and Hutmacher D W 2012 Additive manufacturing of tissues and organs *Prog. Polym. Sci.* **37** 1079–104

[30] Wang X *et al* 2006 Generation of three-dimensional hepatocyte/gelatin structures with rapid prototyping system *Tissue Eng.* **12** 83–90

[31] Chang R, Nam J and Sun W 2008 Effects of dispensing pressure and nozzle diameter on cell survival from solid freeform fabrication-based direct cell writing *Tissue Eng.* A **14** 41–8

[32] Yan Y *et al* 2005 Direct construction of a three-dimensional structure with cells and hydrogel *J. Bioact. Compat. Polym.* **20** 259–69

[33] Xu W *et al* 2007 Rapid prototyping three-dimensional cell/gelatin/fibrinogen constructs for medical regeneration *J. Bioact. Compat. Polym.* **22** 363–77

[34] Fedorovich N E, De Wijn J R, Verbout A J, Alblas J and Dhert W J A 2008 Three-dimensional fiber deposition of cell-laden, viable, patterned constructs for bone tissue printing *Tissue Eng.* **A 14** 127–33

[35] Sellaro T L *et al* 2010 Maintenance of human hepatocyte function *in vitro* by liver-derived extracellular matrix gels *Tissue Eng.* **A 16** 1075–82

[36] Griffith L G and Naughton G 2002 Tissue engineering—current challenges and expanding opportunities *Science* **295** 1009–14

[37] Gaetani R *et al* 2012 Cardiac tissue engineering using tissue printing technology and human cardiac progenitor cells *Biomaterials* **33** 1782–90

[38] Falconnet D, Csucs G, Michelle Grandin H and Textor M 2006 Surface engineering approaches to micropattern surfaces for cell-based assays *Biomaterials* **27** 3044–63

[39] Chang R, Nam J and Sun W 2008 Direct cell writing of 3D microorgan for *in vitro* pharmacokinetic model *Tissue Eng.* **C 14** 157–66

[40] Fischbach C *et al* 2007 Engineering tumors with 3D scaffolds *Nat. Methods* **4** 855–60

[41] Zopf D A, Hollister S J, Nelson M E, Ohye R G and Green G E 2013 Bioresorbable airway splint created with a three-dimensional printer *New Engl. J. Med.* **368** 2043–5

Organ Printing

Dong-Woo Cho, Jung-Seob Lee, Falguni Pati, Jin Woo Jung, Jinah Jang and Jeong Hun Park

Chapter 2

Stereolithography-based 3D printing

3D printing based on selective photo-polymerization, called stereolithography (SL), is the oldest 3D printing technology. SL shows superior performance in the fabrication of 3D structures with very high resolution and accuracy [1, 2]. SL uses ultraviolet (UV) light to induce photo-polymerization, and a liquid photopolymer (known as a photo-curable resin) containing UV light-activated initiator (photo-initiator), monomer and other additives. Controlled UV light is illuminated onto the surface of the liquid photopolymer, and the surface of the liquid layer is spatially solidified by photo-polymerization, which links small molecules (mono-mers) to larger molecules (polymers) composed of many monomer units by triggering free radicals that are produced from the photoinitiator when exposed to UV light of a specific wavelength range. This selective photo-polymerization can be used to fabricate a 2D patterned layer with defined thickness, and a 3D structure can be built by stacking the 2D layers through a layer-by-layer process. After building up the 3D structure, the support platform with the built 3D structure is taken from the vat of liquid photopolymer, and the 3D structure is then washed in a specific alcohol solution to dissolve any remaining liquid photopolymer inside or outside of the structure.

Micro-stereolithography (MSTL) and nano-stereolithography (NSTL) have been developed from SL using specific light systems, so as to build 3D structures at micro- and nanometer scales [3–9], and these technologies have provided an engineering platform for various applications, such as micro-electro-mechanical systems (MEMS) and tissue engineering. In particular, some MSTL and NSTL systems have been developed and used in the fabrication of biomaterial matrices (known as scaffolds) and cell-laden structures in the tissue engineering field [10–17]. This chapter describes the general configurations of SL systems, and their advantages and disadvantages are also discussed.

doi:10.1088/978-1-6817-4079-9ch2   2-1

2.1 System configurations

In SL systems, a 3D structure, consisting of multiple solidified 2D patterned layers, is built within the z-axis support platform using a layer-by-layer process. SL systems can be classified into two kinds, according to their construction orientation: bottom-up and top-down.

In a bottom-up system, the first layer is solidified in a layer of the liquid photopolymer and attached onto the support platform through controlled UV light illumination from the top. The support platform that the solidified layer is attached to then lowers into the vat of liquid photopolymer from the surface to allow the liquid photopolymer to stratify the surface of the solidified layer with defined thickness. The next layer is solidified in a layer of liquid photopolymer and adhered to the upper surface of the previously solidified layer. These steps (the movement of the z-axis support platform, the layer stratification of liquid photopolymer and the 2D layer solidification in a layer of liquid photopolymer) are repeated to complete the fabrication of the 3D structure.

This system uses a free surface condition in which the surface of the liquid photopolymer is free and in contact with air (figure 2.1(A)). In the free surface condition, it is difficult to control the thickness of each liquid photopolymer layer precisely because of the viscosity of the liquid photopolymer, and a relatively long time is required for stabilization. A 'wiper arm' can be used to flatten the surface of the newly stratified liquid photopolymer; however, this kind of fast formation of the new liquid photopolymer layer can include unintended air bubbles due to the surface tension of the liquid photopolymer.

A top-down system uses the constraint surface condition in which the surface of the liquid photopolymer is constrained by a transparent window (figure 2.1(B)). Controlled UV light is illuminated from the bottom and transferred to the liquid photopolymer layer through the transparent window. The first layer is solidified in a

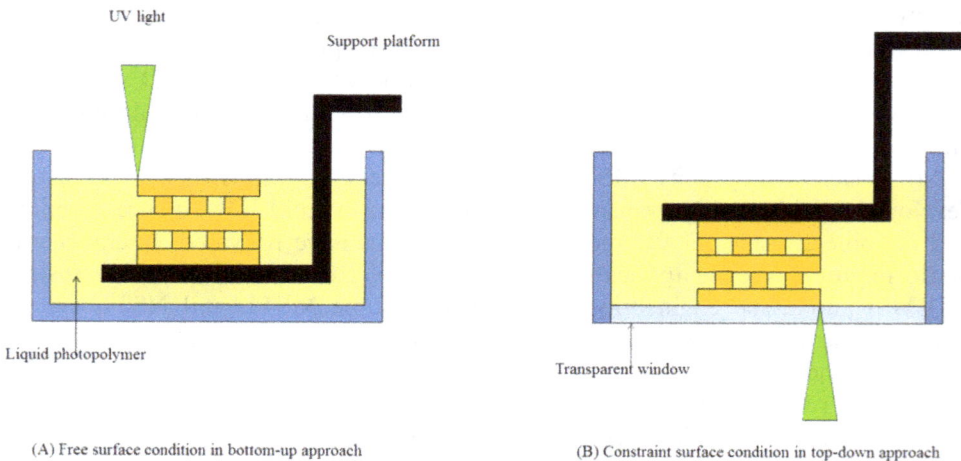

(A) Free surface condition in bottom-up approach   (B) Constraint surface condition in top-down approach

Figure 2.1. Schematic illustrations of system set-ups for the SL process according to construction orientation: (A) free surface condition in a bottom-up system and (B) constraint surface condition in a top-down system.

layer of liquid photopolymer between the transparent window and the bottom surface of the support platform, and then separated from the surface of the transparent window by raising the support platform with a defined thickness. The next pattern is then solidified in a layer of liquid photopolymer between the transparent window and bottom surface of the previously solidified pattern layer. These steps (the movement of the z-axis support platform, the layer stratification of liquid photopolymer and the 2D pattern solidification in a layer of liquid photo-polymer) are also then repeated to complete the fabrication of the 3D structure.

In the constraint surface condition in a top-down system, the stratification of new liquid photopolymer on the previously solidified layer is not required and the surface of the pattern layer that is being solidified is not exposed to the air; therefore, the effects of oxygen inhibition are insignificant.

In addition, this surface condition can allow more accurate control of the layer thickness of the newly stratified liquid photopolymer by controlling the height of a transparent window, and completion of the process in a short time. However, the solidified pattern can sometimes adhere to the surface of the transparent window, which can result in partial or total destruction of the structure during the construction process. In this regard, the transparent window should have a hydro-phobic surface for easy detachment from the solidified pattern. Several other requirements must also be considered for the transparent window in the constraint surface condition: (a) high transmittance for the UV wavelength range, (b) a durable and fine surface and (c) chemical stability in the photo-polymerization process.

Additionally, SL systems can be classified into three types according to the 2D patterning process for one layer fabrication: beam-scanning-, image-projection- and two-photon-based systems.

Beam-scanning-based SL was the first version to use the scanning of a UV laser beam to draw a pattern. Most SL systems in use have the same working mechanism as this laser beam scanning system. In the system, a UV laser beam is focused by specific optics, including a focusing lens, and scanned onto the surface of the liquid photopolymer by x–y stage movement or a dynamic focusing lens, controlled by a computer (figure 2.2(A)). It is generally difficult to maintain the focal point of the light beam over the planar surface of the liquid photopolymer using a dynamic light focusing lens, and UV laser beam scanning by x–y stage movement can achieve a higher resolution. The layer of the liquid photopolymer is solidified along the

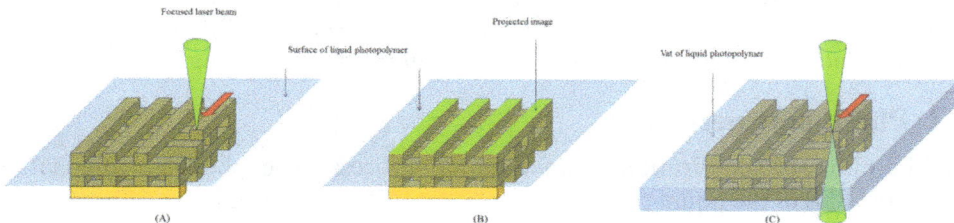

Figure 2.2. Schematic illustrations of SL system set-ups according to the 2D patterning process: (A) beam scanning, (B) image projection and (C) two-photon-based SL systems.

moving path of the scanned laser beam, and a 2D pattern layer is formed according to the scanned area of the UV laser beam.

The beam-scanning-based method usually has a higher resolution than image-projection-based systems. However, employing a focused laser beam can be too slow to fabricate a 3D structure of a large size because too much scanning of the laser beam is required to solidify a single 2D patterned layer. A laser-beam-scanning-based SL system using an optic fiber array was developed to fabricate multiple 3D structures in a single process [18]; however, movement of the support platform in the x–y–z directions in the vat of the liquid photopolymer can still result in the collapse of the structure in the liquid photopolymer.

Next, an image-projection-based SL system was developed. This uses an image generation device, such as a liquid crystal display (LCD) or digital micro-mirror device (DMD), and a lamp instead of a laser beam. A 2D binary image is generated from an image generation device and illuminated on the surface of the liquid photopolymer (figure 2.2(B)). Generally, a DMD, an array of several million micro-mirrors that can be rotated independently to an on or off state, has a relatively high reflectance over a wide range of wavelengths, while a LCD has lower transmittance in the range of UV wavelength. A DMD also has a higher switching speed and a smaller pixel size than a LCD.

In this system, the entire 2D pattern is solidified by a single illumination of the 2D pattern image onto the surface of liquid photopolymer, resulting in a significant decrease in fabrication time compared with a beam-scanning-based system. In addition, the surface of the 2D layer solidified by image illumination is always flat and smooth, whereas the surface is uneven in the laser-beam-scanning-based system.

A two-photon-based SL is based on simultaneous two-photon absorption within very small regions. When the liquid photopolymer absorbs two photons, each with the same wavelength at the same time and point, the two photons act as one photon of a wavelength that is two times higher than the one photon, and the liquid photopolymer can be polymerized and solidified (figure 2.2(C)). Movement of the focal point inside of the liquid photopolymer then enables fabrication of a 3D structure without the use of a support platform in a vat of liquid photopolymer.

In two-photon-based SL, the solidification happens only at the very narrow focal point where the light intensity is highest, and other regions are not affected inside the liquid photopolymer. The resolution of spatial solidification that can be achieved by two-photon-based SL is almost 100 nm, which is below the diffraction limit of the applied light beam. This fascinating advantage gives two-photon-based SL its superior accuracy and resolution, better than those for all other 3D printing technologies.

2.2 Photo-polymerization

In SL, photo-polymerization is an essential process. The photo-polymerization characteristics of the liquid photopolymer used in SL are critical in the fabrication of the 3D structure, because it has significant effects on the solidification time, resolution and thickness of the 2D patterned layer that is solidified.

In SL systems, photo-polymerization of the liquid photopolymer can be controlled by adjusting the exposure conditions [19]. The solidified thickness and width of the liquid photopolymer depend primarily on the power and scanning speed of the laser beam (for laser-beam-scanning-based and two-photon-based systems) and the power of the light lamp and the illumination time of the projected image (for image-projection-based systems). This can also be controlled by the chemistry and the quantity of monomers and photoinitiators in the liquid photopolymer. UV absorbers are occasionally added to the liquid photopolymer to control the layer thickness.

References

[1] Melchels F P W, Feijen J and Grijpma D W 2010 A review on stereolithography and its applications in biomedical engineering *Biomaterials* **31** 6121–30

[2] Seol Y J, Kang T Y and Cho D W 2012 Solid freeform fabrication technology applied to tissue engineering with various biomaterials *Soft Matter.* **8** 1730–35

[3] Lee I H and Cho D W 2003 Micro-stereolithography photopolymer solidification patterns for various laser beam exposure conditions *Int. J. Adv. Manuf. Technol.* **22** 410–16

[4] Maruo S and Ikuta K 2002 Submicron stereolithography for the production of freely movable mechanisms by using single-photon polymerization *Sensors and Actuators* A **100** 70–6

[5] Sun C, Fang N, Wu D M and Zhang X 2005 Projection micro-stereolithography using digital micro-mirror dynamic mask *Sensors and Actuators* A **121** 113–20

[6] Liska R *et al* 2007 Photopolymers for rapid prototyping *J. Coat Technol. Res.* **4** 505–10

[7] Xing J F *et al* 2007 Improving spatial resolution of two-photon microfabrication by using photoinitiator with high initiating efficiency *Appl. Phys. Lett.* **90** 131106

[8] Choi J S, Kang H W, Lee I H, Ko T J and Cho D W 2009 Development of micro-stereolithography technology using a UV lamp and optical fiber *Int. J. Adv. Manuf. Technol.* **41** 281–89

[9] Zheng X, Deotte J, Alonso M P, Farquar G R, Weisgraber T H, Gemberling S, Lee H, Fang N and Spadaccini C M 2012 Design and optimization of a light-emitting diode projection micro-stereolithography three-dimensional manufacturing system *Rev. Sci. Instrum.* **83** 125001

[10] Melchels F P W, Feijen J and Grijpma D W 2009 A poly(D,L-lactide) resin for the preparation of tissue engineering scaffolds by stereolithography *Biomaterials* **30** 3801

[11] Melchels F P W, Bertoldi K, Gabbrielli R, Velders A H, Feijen J and Grijpm D W 2010 Mathematically defined tissue engineering scaffold architectures prepared by stereolithography *Biomaterials* **31** 6909

[12] Seck T M, Melchels F P W, Feijen J and Grijpma D W 2010 Designed biodegradable hydrogel structures prepared by stereolithography using poly(ethylene glycol)/poly (D,L-lactide)-based resins *J. Control. Release* **148** 34–41

[13] Lee J W, Kang K S, Lee S H, Kim J Y, Lee B K and Cho D W 2011 Bone regeneration using a microstereolithography-produced customized poly(propylene fumarate)/diethyl fumarate photopolymer 3D scaffold incorporating BMP-2 loaded PLGA microspheres *Biomaterials* **32** 744–52

[14] Kang H W, Park J H, Kang T H, Seol Y J and Cho D W 2012 Unit cell-based computer-aided manufacturing system for tissue engineering *Biofabrication* **4** 015005

[15] Kang H W and Cho D W 2012 Development of an indirect stereolithography technology for scaffold fabrication with a wide range of biomaterial selectivity *Tissue Eng.* **18** 719–29

[16] Lin H, Zhang D, Alexander P G, Yang G, Tan J, Cheng A W M and Tuan R S 2013 Application of visible light-based projection stereolithography for live cell-scaffold fabrication with designed architecture *Biomaterials* **34** 331–9

[17] Park J H *et al* 2015 A novel tissue-engineered trachea with a mechanical behavior similar to native trachea *Biomaterials* **62** 106–15

[18] Ikuta K, Ogata T, Tsubio M and Kojima S 1996 Development of mass productive micro stereo lithography (Mass–IH process) *Proc. IEEE Micro Electro Mech. Syst.* 301–6

[19] Lee I H and Cho D W 2004 An investigation on photopolymer solidification considering laser irradiation energy in micro-stereolithography *Microsyst. Technol.* **10** 592–8

Dong-Woo Cho, Jung-Seob Lee, Falguni Pati, Jin Woo Jung, Jinah Jang and Jeong Hun Park

Chapter 3

Inkjet-based 3D printing

Woodblock printing technology was developed many years ago to produce paper copies rapidly. Today, inkjet printing machines have largely replaced the traditional printing machine, and have become popular for distributing and sharing information in offices and households. This technique has been widely used in various industrial fields. For example, date and product information are printed onto cans and bottles [1]. Also, polymer or ceramic electronics, such as polymer light-emitting diodes (PLED) [2] and solid oxide fuel cells (SOFC) [3], can be printed with inkjet printing systems.

The inkjet printing technique evolved into 3D inkjet printing, whereby a 3D object is fabricated by stacking sequential layers of cross-sectional slices. While a 2D inkjet printing system prints an image copy, a 3D inkjet printing system prints a 3D structure, based on a digital 3D model, which is designed using CAD software or developed with a 3D scanner. A 3D inkjet printing system offers an opportunity to take advantage of the processing of high-resolution 2D-patterned arrays via droplet ejection through the inkjet-head nozzle. For organ printing, this advantage allows the powerful direct-patterning of cells at desired locations, because the printing droplet size can be regulated at the single-cell level [4]. Thus, direct cell patterning with a 3D inkjet printing system could be effective for organizing a pre-tissue with spatially organized cells. Additionally, the operating system for multi-color printing not only enables the production of a construct that is composed of a few types of cells, it also enables the production of one containing gradients in cell composition for interface tissue engineering (ITF) [5]. In this chapter, we explore the principles of 3D inkjet printing, describe how it is being used for tissue engineering applications and discuss its limitations and future directions.

3.1 Inkjet printing technology

Inkjet printing techniques are classified into drop-on-demand (DOD) and continuous types (figure 3.1) [6]. The continuous inkjet (CIJ) printing technique was first commercialized in 1951, by Siemens. In this system, a continuous stream of ink, generated by a

3-1

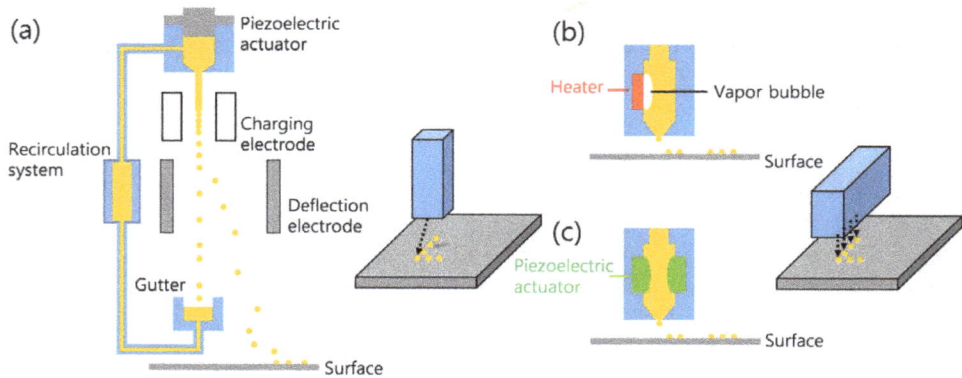

Figure 3.1. Schematics of inkjet printing systems. (a) Continuous inkjet printing: a multiple-deflection system, (b) thermal drop-on-demand inkjet printing and (c) piezoelectric drop-on-demand inkjet printing.

high-pressure pump, is stimulated by a high-frequency acoustic wave from a piezo-electric device to form droplets with consistent volume at regular intervals. The droplets pass through an electrostatic field, generated by a charge electrode, to apply, selectively, an electrical charge to a droplet. The path of each charged droplet is then deflected by an electrostatic deflector plate to form a desired pattern on a substrate. Uncharged droplets move straight towards a gutter for reuse. However, due to this complex system for droplet path control and ink recycling, CIJ printing is used mainly in industry. DOD inkjet printing places droplets directly at an exact point. According to a digital signal, the droplet can be generated selectively by a pressure change in the ink reservoir, resulting from a piezoelectric effect (piezoelectric DOD inkjet printing) or temporal bubble formation (thermal DOD inkjet printing). Because such DOD-based inkjet printing equipment is usually lighter, smaller and cheaper than that for CIJ, it is used more widely in households, offices and industry.

Inkjet printing (CIJ and DOD) can deposit metal [7, 8], ceramic [9, 10] or polymer [11, 12] materials in the liquid state as well as liquid ink [13]. Normally, any solid material is processed to form a fine powder and then dispersed uniformly in a liquid [14]. Alternatively, a soluble material can be dissolved in a specific solvent. This solution or suspension should have rheological properties and a surface tension suitable for the inkjet printing mechanism [15]. As described in previous studies, the fluidic behavior of droplet formation can be predicted by the Weber number (We), the Reynolds number (Re) and the Ohnesorge number (Oh) [13, 15, 16], defined by

$$\mathrm{Re} = \frac{\rho D V}{\eta}, \qquad \mathrm{We} = \frac{\rho D V^2}{\sigma}, \qquad \mathrm{Oh} = \frac{\sqrt{\mathrm{We}}}{\mathrm{Re}} = \frac{\eta}{\sqrt{\rho \sigma D}},$$

where η, ρ, σ and D are the dynamic viscosity, density, surface tension and characteristics length, respectively. The Ohnesorge number should be in the range of 0.1–1 for consistent droplet formation and ejection [15, 17, 18]. In the range Oh < 0.1, multi-droplets are formed by one jetting, and for too high an Ohnesorge number no droplet is generated due to viscous dissipation [19, 20].

Thermal DOD inkjet printing systems are used in most desktop or domestic printers, because of their simplicity, low cost and miniaturization. Piezoelectric DOD inkjet printing has been preferred for the printing of various materials in industrial manufacturing processes [13, 18, 21] for two main reasons: (1) the spacing and velocity of a droplet can be controlled readily by changing the actuation pulse generated by the piezoelectric device, and (2) the heater in a thermal DOD inkjet printing system can influence the properties of a nanoparticle. CIJ printing has also been used successfully in the industrial field to build objects from various materials. The droplet ejection rate of CIJ printing is much faster than that of DOD printing. However, CIJ printing has the drawback of contaminating the ink provided through the recycling system [13, 14].

3.2 3D inkjet printing technology

In the 1980s, SL was introduced as the first 3D printing (or additive manufacturing) technology [22]. The 3D printing system created a 3D object based on a successive stacking process of UV-cured layers. In 1989, this layer-by-layer process was applied to an inkjet printing system to enable rapid production of a 3D structure [23, 24]. In this system, layers of material are printed pixel-wise (or point-by-point) using multi-colored tiny droplets generated from nozzles within the inkjet head [25]. The volume of the droplet is influenced by the viscosity and surface tension of solutions, and the pattern and magnitude of the pressure wave within the nozzle system. The tiny droplets generated under suitable droplet ejecting conditions are dispensed onto a planar surface to create a thin layer with a specific pattern, which represents a cross-section of a 3D object. Generally, the printed layer in the liquid state should then be solidified by a chemical or physical treatment. Then, another layer is created on the first layer in the same way. A 3D object is fabricated by repetition of these steps in a layer-by-layer manner.

There are two 3D inkjet printing approaches for stacking layers: powder-bed-based and direct inkjet printing techniques (figure 3.2). In the powder-bed-based printing process [1], powder particles and a binder or solvent are used to pattern the cross-section of a layer. A binder or solvent ejected from an inkjet printer head infiltrates the tiny powders in a powder bed to bond them to each other. For patterning the next 2D image, powders cover the printed layer to form a new powder layer. This process is repeated for each layer and the unprinted powder bed plays a role as a self-supporting part to create overhang, undercut or other complex shapes for the object [26]. At the end of the printing process, the 3D object is obtained by removing the residual non-bonded powder.

By contrast, direct inkjet printing dispenses a sequence of liquid droplets onto a plane to form a pattern of layers. The pattern of the liquid material is then solidified by solvent evaporation, polymer crosslinking, sintering, crystallization or vitrification [15, 17, 27]. However, for fabrication of a scaffold, overhang or undercut structure, it may require a temporary supporting part, which could be generated by a second inkjet nozzle.

Figure 3.2. Schematics of 3D inkjet printing. (a) Powder-bed-based inkjet printing and (b) direct inkjet printing.

3.3 3D inkjet printing technique for creating 3D functional tissue/organ constructs

Both powder-bed-based and direct 3D inkjet printing techniques are capable of creating 3D scaffolds using various types of biocompatible materials [28, 29]. In particular, 3D inkjet printing is effective in providing precise control for the placement of cells and biological hydrogels. Because many tissues and organs are generally composed of heterogeneous tissue, it may be necessary to situate biomaterials and living cells in desired positions for effective regeneration of a functional tissue/organ. The inkjet printing head can eject tiny droplets of inks of several colors to form a 2D pattern with high resolution (under ~20 µm) on demand [30]. Thus, this technology has the potential to position individual droplets using single cells and is applicable for positioning multiple bio-components with living cells under digital data for a target tissue/organ. In addition, the 3D inkjet printing process can be applied to building 3D pre-tissues or pre-organs derived from biomaterials and living cells.

Cell positioning using the inkjet printing system was first introduced by Klebe [31] in 1988. Pre-inkjet-printed patterns of cell adhesion proteins on the plate enabled the spatially selective attachment of cells. In 2003, direct inkjet printing of cells was proposed by Boland *et al* for patterning of mammalian cells [32, 33]. The cells were suspended in culture medium and placed inside a thermal DOD inkjet head. A 2D layer of a circular pattern was then printed onto a culture plate. More than 90% viability of cells was observed in the printed pattern. In 2005, Nakamura *et al* introduced a high-resolution cell-positioning technique based on a static-electricity-actuated inkjet (SEAJet) head, which ejects an ink droplet by changing the volume of the ink cavity, in a similar way to piezoelectric DOD inkjet printing [30]. They presented a precise 2D dot-pattern containing 0–4 cells per dot. The printed dot diameter could be controlled from 85 to 240 µm.

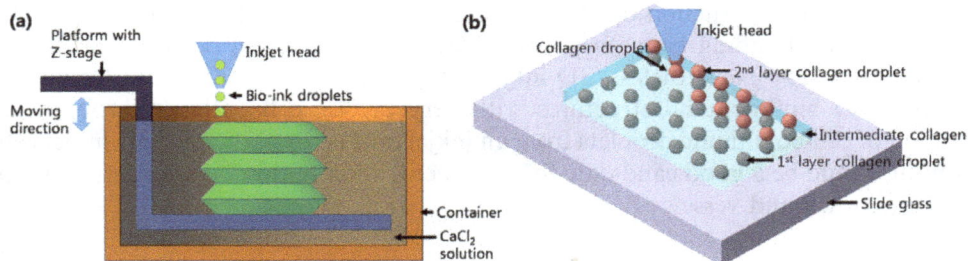

Figure 3.3. Schematics of 3D inkjet cell printing. (a) Platform-assisted 3D inkjet bioprinting system [35] and (b) printing platform for 3D cell-laden droplet printing [36].

Through successive stacking of such 2D cell patterns, 3D structures with living cells can be constructed. In 2009, Nakamura *et al* demonstrated a 3D bioprinter that formed a 3D cell-supporting structure using an inkjet technique and hydrogels [34]. The bioink was prepared by encapsulation of HeLa cells within sodium alginate solution, and then printed onto a reservoir of calcium chloride solution. Because the droplets of sodium alginate are rapidly crosslinked in calcium chloride solution, the alginate gel beads had almost the same size and shape as the droplets. The patterned droplets stayed stably at the initially ejected position because of a highly viscous calcium chloride solution with polyvinyl alcohol. The pattern of the next layer was successively printed on the crosslinked pattern, to obtain a 3D gel structure. Xu *et al* presented a more advanced 3D cell inkjet printing system, containing a Z-shaped platform attached to a z-stage within a reservoir of calcium chloride solution (figure 3.3(a)) [35]. This platform was effective in correctly designating the printing plane for each layer. This platform-assisted 3D inkjet bioprinting system enabled the construction of a zigzag tube structure composed of fibroblasts. This printing of the hydrogel droplet containing cells in a reservoir requires the rapid crosslinking and stable positioning of the droplet. However, this process is not suitable for locating a slowly crosslinked droplet stably, because the droplet can be mixed with the reservoir solution before its crosslinking. Moon *et al* proposed a cell-laden hydrogel droplet deposition system for printing 3D structures composed of cell-encapsulated collagen (figure 3.3(b)) [36]. The cell/collagen suspension was printed on a substrate at room temperature, and the printed pattern was subsequently crosslinked at 37 °C for 5 min. By repeating this process of layering, it was possible to make a 0°/90° crossover cell pattern composed of multiple cells.

3.4 Conclusion

In this chapter, the background to and the current state of 3D inkjet printing for tissue engineering have been described. 3D inkjet printing is a versatile method that enables a variety of materials to be constructed into a 3D scaffold, in addition to facilitating high-precision cell patterning within a 3D hydrogel construct. However, from an engineering viewpoint, several issues should be mentioned, such as resolution and mechanical properties. Compared with other 3D printing techniques, the construction of 3D structures using inkjet printing can lead to low strength, weak bonding between powder particles and trapped powders inside the created body [28]. Nevertheless, 3D inkjet cell printing has significant potential for creating 3D biofunctional structures with very precise cell patterns through the ejection of cell-containing droplets from an inkjet nozzle. In particular, 3D inkjet cell printing would be a good candidate for the regeneration of thin-layered tissues, such as skin, cornea and vessels.

References

[1] Calvert P 2001 Inkjet printing for materials and devices *Chem. Mater.* **13** 3299–305

[2] Dijksman J F *et al* 2007 Precision ink jet printing of polymer light emitting displays *J. Mater. Chem.* **17** 511–22

[3] Tomov R I *et al* 2010 Direct ceramic inkjet printing of yttria-stabilized zirconia electrolyte layers for anode-supported solid oxide fuel cells *J. Power Sources* **195** 7160–7

[4] Barry R A, Shepherd R F, Hanson J N, Nuzzo R G, Wiltzius P and Lewis J A 2009 Direct-write assembly of 3D hydrogel scaffolds for guided cell growth *Adv. Mater.* **21** 2407

[5] Seidi A, Ramalingam M, Elloumi-Hannachi I, Ostrovidov S and Khademhosseini A 2011 Gradient biomaterials for soft-to-hard interface tissue engineering *Acta Biomater.* **7** 1441–51

[6] Derby B 2008 Bioprinting: inkjet printing proteins and hybrid cell-containing materials and structures *J. Mater. Chem.* **18** 5717–21

[7] Ko S H, Chung J, Hotz N, Nam K H and Grigoropoulos C P 2010 Metal nanoparticle direct inkjet printing for low-temperature 3D micro metal structure fabrication *J. Micromech. Microeng.* **2010** 20

[8] Ishida Y, Nakagawa G and Asano T 2007 Inkjet printing of nickel nanosized particles for metal-induced crystallization of amorphous silicon *Japan J. Appl. Phys.* **46** 6437–43

[9] Xiang Q F, Evans J R G, Edirisinghe M J and Blazdell P F 1997 Solid freeforming of ceramics using a drop-on-demand jet printer (*Proc. Inst. Mech. Eng. B*) *J. Eng. Manufacture* **211** 211–4

[10] Slade C E 1998 Freeforming ceramics using a thermal jet printer *J. Mater. Sci. Lett.* **17** 1669–71

[11] Jang D, Kim D and Moon J 2009 Influence of fluid physical properties on ink-jet printability *Langmuir* **25** 2629–35

[12] Wang J Z, Zheng Z H, Li H W, Huck W T S and Sirringhaus H 2004 Dewetting of conducting polymer inkjet droplets on patterned surfaces *Nat. Mater.* **3** 171–6

[13] Derby B 2010 Inkjet printing of functional and structural materials: fluid property requirements, feature stability and resolution *Annu. Rev. Mater. Res.* **40** 395–414

[14] Derby B and Reis N 2003 Inkjet printing of highly loaded particulate suspensions *MRS Bull.* **28** 815–8

[15] Hon K K B, Li L and Hutchings I M 2008 Direct writing technology—advances and developments *CIRP Annals—Manufacturing Technology* **57** 601–20

[16] Ozkol E, Ebert J, Uibel K, Watjen A M and Telle R 2009 Development of high solid content aqueous 3Y-TZP suspensions for direct inkjet printing using a thermal inkjet printer *J. Eur. Ceram. Soc.* **29** 403–9

[17] Noguera R, Lejeune M and Chartier T 2005 3D fine scale ceramic components formed by ink-jet prototyping process *J. Eur. Ceram. Soc.* **25** 2055–9

[18] Derby B 2011 Inkjet printing ceramics: from drops to solid *J. Eur. Ceram. Soc.* **31** 2543–50

[19] Lee A, Sudau K, Ahn K H, Lee S J and Willenbacher N 2012 Optimization of experimental parameters to suppress nozzle clogging in inkjet printing *Ind. Eng. Chem. Res.* **51** 13195–204

[20] Shen W F, Zhang X P, Huang Q J, Xu Q S and Song W J 2014 Preparation of solid silver nanoparticles for inkjet printed flexible electronics with high conductivity *Nanoscale* **6** 1622–8

[21] Kim C S, Park S J, Sim W, Kim Y J and Yoo Y 2009 Modeling and characterization of an industrial inkjet head for micro-patterning on printed circuit boards *Comput. Fluids* **38** 602–12

[22] Ventola C L 2014 Medical applications for 3D printing: current and projected uses *Pharmacy and Therapeutics* **39** 704–11 PMCID: PMC4189697

[23] Houben R J 2012 Equipment for printing of high viscosity liquids and molten metals *University of Twente* PhD Thesis available here http://doc.utwente.nl/81666/

[24] Pfister A, Landers R, Laib A, Hubner U, Schmelzeisen R and Mulhaupt R 2004 Biofunctional rapid prototyping for tissue-engineering applications: 3D bioplotting versus 3D printing *J. Polym. Sci. Pol. Chem.* **42** 624–38

[25] Levi B G 2001 New printing technologies raise hopes for cheap plastic electronics *Phys. Today* **54** 20–2

[26] Hong S B, Eliaz N, Sachs E M, Allen S M and Latanision R M 2001 Corrosion behavior of advanced titanium-based alloys made by three-dimensional printing (3DPTM) for biomedical applications *Corros. Sci.* **43** 1781–91

[27] Singh M, Haverinen H M, Dhagat P and Jabbour G E 2010 Inkjet printing—process and its applications *Adv. Mater.* **22** 673–85

[28] Butscher A, Bohner M, Hofmann S, Gauckler L and Muller R 2011 Structural and material approaches to bone tissue engineering in powder-based three-dimensional printing *Acta Biomater.* **7** 907–20

[29] Boland T *et al* 2007 Drop-on-demand printing of cells and materials for designer tissue constructs *Mat. Sci. Eng.* C **27** 372–6

[30] Nakamura M *et al* 2005 Biocompatible inkjet printing technique for designed seeding of individual living cells *Tissue Eng.* **11** 1658–66

[31] Klebe R J 1988 Cytoscribing: a method for micropositioning cells and the construction of two- and three-dimensional synthetic tissues *Exp. Cell Res.* **179** 362–73

[32] Xu T, Jin J, Gregory C, Hickman J J and Boland T 2005 Inkjet printing of viable mammalian cells *Biomaterials* **26** 93–9

[33] Wilson W C Jr and Boland T 2003 Cell and organ printing 1: protein and cell printers *Anat. Rec.* A **272** 491–6

[34] Nishiyama Y *et al* 2009 Development of a three-dimensional bioprinter: construction of cell supporting structures using hydrogel and state-of-the-art inkjet technology *J. Biomech. Eng.* **2009** 131

[35] Xu C, Chai W, Huang Y and Markwald R R 2012 Scaffold-free inkjet printing of three-dimensional zigzag cellular tubes *Biotechnol. Bioeng.* **109** 3152–60

[36] Moon S *et al* 2010 Layer by layer three-dimensional tissue epitaxy by cell-laden hydrogel droplets *Tissue Eng.* C **16** 157–66

Organ Printing

Dong-Woo Cho, Jung-Seob Lee, Falguni Pati, Jin Woo Jung, Jinah Jang and Jeong Hun Park

Chapter 4

Dispensing-based 3D printing

3D printing technology (based on dispensing to fabricate and regenerate target tissue and organs) originated from fused deposition modeling (FDM), a rapid prototyping technology. FDM was developed at the end of the 1980s and commercialized in the early 1990s as a 3D printer [1]. The process of FDM is as follows: (1) A material line, like a filament, is extruded from a nozzle with a dispensing system; (2) a 2D pattern is simultaneously fabricated by movement of the dispensing head, which extrudes the line or a bed where the structure is stacked up; and (3) through a layer-by-layer process, the desired 3D structure is finally fabricated [2, 3]. FDM is a bottom-up system and has the advantage of directly fabricating the desired 3D structure; it also has a relatively fast fabrication time, because there is no post-processing.

The main materials in FDM are thermoplastic, and these are stable during the melting process [4]. Thermoplastic biomaterials, such as poly-caprolactone (PCL), were developed at the end of the 1990s. FDM is capable of fabricating scaffolds that can be implanted in the human body for tissue and organ reconstruction or regeneration, without severe side-effects [5–7]. Thus, at the end of the 1990s much research was conducted in the tissue engineering field with dispensing-based 3D printing. In the 2000s, many researchers fabricated 3D porous scaffolds with dispensing-based 3D printing to regenerate and reconstruct bone and cartilage. At the end of the 2000s, studies directly printing various kinds of cells or cell-laden hydrogels as well as thermoplastic biomaterials were reported and they regenerated relatively simple organs, such as the bladder, air pathway and skin [8–10].

More recently, novel research has been conducted to fabricate a tissue-engineered structure with cell printing technology and to validate the inherent function of the tissue-engineered structure for the regeneration of more complex organs, such as the heart and liver. Dispensing-based 3D printing, from the original FDM, has used various systems, depending on the objective of the target regeneration. In this chapter, we introduce a general 3D printing system based on dispensing with

doi:10.1088/978-1-6817-4079-9ch4

biomaterials, discussing its advantages and disadvantages, and we explain organ-printing technology [11].

4.1 System configurations

In dispensing-based 3D printing, the 3D structure is fabricated directly with material extruded through the nozzle, unlike in 3D printing based on a laser. The fabrication process of a 3D structure is classified into three steps: material preparation, code generation and fabrication with the 3D printing system. First, biomaterial (as a prepared filament or pellet-formed material) is put into the syringe or dispensing equipment (figure 4.1(a) and (b)). Code (including beam path information for the 3D structure's shape) is then generated using CAM software, which can convert the STL format file of a 3D CAD model to code. In the fabrication procedure, the molten material in the syringe or organic solvent is extruded from the nozzle by the dispenser, and the dispensed line makes the desired 2D pattern with a pre-programmed beam path. The 2D pattern is stacked up at a sub-hundred micrometer height, in a layer-by-layer process, to fabricate the final 3D structure.

For 3D structure fabrication, a system composed of a 3D motion stage, enabling very precise and accurate control at the sub-micrometer scale, and a dispensing system controlling the amount of material flow in micro- and nanoliters, is set up. The 3D motion stage enables head movement to effect precise and accurate positioning along the x, y and z axes (Cartesian coordinates) in the dispensing of the molten material. Representative 3D motion stages are of two types: a 'gantry stage' (figure 4.1(c)), with movement of the head, including dispensing and heating systems in aggregate, without bed movement, and a 'stacked stage', with movement of the head in the z-axis and the bed in the x–y axis. Based on these two stage types,

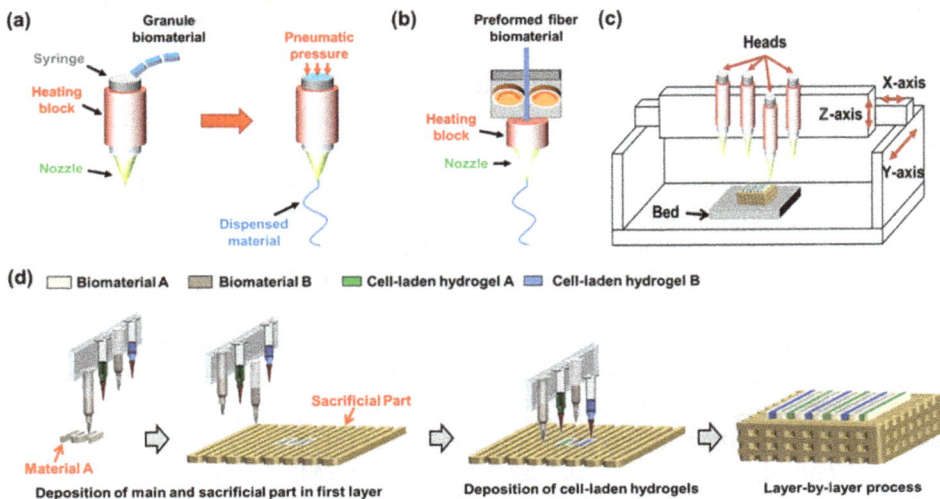

Figure 4.1. Schematic illustrations of dispensing-based 3D printing. (a) Syringe-based dispensing system, (b) typical FDM dispensing system, (c) 3D motion stage (gantry type) and (d) structure fabrication method.

various modified 3D printing systems have been developed in line with their different research purposes. The dispensing system has heating and/or cooling blocks and a dispenser to control the minimum material flow rate with pneumatic pressure, pistons and screws. Selection of the three kinds of dispensers depends on the viscosity and minimum flow rate of the materials used. Nearly 30 000 3D printers are sold worldwide every year, and academic institutions are increasingly purchasing and applying dispensing-based 3D printing technology in tissue and organ engineering research [12]. A few systems use multiple heads to dispense several materials without retooling (figure 4.1(d)) [6, 13–15].

4.2 Organ printing based on dispensing

The advantages of dispensing-based 3D printing over other systems are ease of use, simple drive-mechanism components and the absence of post-processing requirements. However, the drawbacks of the FDM system and 3D dispensed printing are its need for preformed fibers with uniform size and material properties, relatively slow fabrication time compared with inkjet printers and reduced maximum force capability due to pneumatic pressure.

Many researchers in the tissue engineering field have reformed PCL materials to filament shape or modified the typical FDM system. Hutmacher, Teoh, Zein and colleagues [16–18] extruded PCL in a filament shape in preparing the material, because PCL is a flexible, bioresorbable polymer with good fabrication fidelity, and they used the PCL filaments in a commercial FDM system (Stratasys Inc., MN, USA). They fabricated a porous scaffold with PCL and conducted performance tests in terms of mechanical properties and cell viability for bone and cartilage regeneration.

Others renovated a typical FDM system or developed new dispensing systems to reduce the preparation of PCL filaments. They remodeled the dispensing part of the FDM system with syringes, enabling PCL solid pellets, or developed new systems with multi-heads that included syringes and heating blocks. Various scaffolds with biomaterials were adequately fabricated for target tissues and organs using the modified systems, and they seeded cells into scaffolds and evaluated tissue formation.

Sun et al [6, 9] developed the 'precision extruding deposition' (PED) system. In contrast to the typical FDM process with the precursor filaments, the PED process extrudes granulated or pellet biomaterials directly without filament preparation, and dispenses the PCL to fabricate a porous scaffold with constant pore size using CAD/CAM.

Woodfield et al developed the 'fiber deposition technique', using a syringe and plunger dispenser, and fabricated a co-polymer scaffold with an interconnecting pore network using poly(ethylene glycol)-terephthalate and poly(butylene terephthalate) blocks for articular cartilage regeneration [13].

Cho et al developed a multi-head deposition system (MHDS) with four dispensing heads, including syringes, a heating system and pneumatic pressure [14]. They studied various kinds of tissue and organ regeneration, as well as bone and cartilage regeneration, with synthetic (PCL, PLGA, blended PCL/PLGA, TCP (tricalcium phosphate) and PEG and natural (collagen, alginate, agarose and dECM (decellularized extra cellular matrix)) polymers [15, 19]. Shim *et al* fabricated a thin scaffold with blended PCL/PLGA/TCP with rh-BMP-2 for regeneration of the calvarial bone of rabbits [20]. They also fabricated a PCL/PLGA/collagen/rh-BMP-2 porous hybrid structure similar to a long bone defect in rabbits and evaluated osteogenesis [21].

In these various examples, systems with multi-heads are capable of fabricating one structure with different materials. These systems can fabricate the desired 3D structure with biomaterials, and can also provide a cell-printed structure with cells, cell spheroids and cell-laden hydrogels [7, 22]. For more effective tissue and organ regeneration, cells or cell-laden hydrogels, with cells encapsulated in hydrogels at high densities, can be dispensed directly in accurate positions with good viability using a dispensing-based 3D printing system. The dispensed cells are responsible for tissue and organ formation through cell-to-cell or cell-to-ECM combinations. This method, called 3D cell printing technology, has contributed to improvements in tissue regeneration in tissue engineering and regenerative medicine.

Mironov, Forgacs, and colleagues [7] developed directly dispensing cells, or cell spheroid technology, using a commercial system, a 'dispensing laboratory bioprinter' (LAB) and a 3D robotic printer for enhanced tissue regeneration in, for example, vascular networks. Schuurman, Fedorovich, Malda *et al* [10, 23] developed a cell-printed structure, which consisted of PCL and chondrocyte-laden alginate hydrogel, using a commercial Bioscaffolder dispensing system (SYS+ENG, Salzgitter-Bad, Germany) to form bone, cartilage, and osteochondral tissue. The Bioscaffolder dispensing system has two dispensing heads with a heating block to dispense molten PCL and the cell-laden hydrogel. Cho *et al* [15] developed the multi-head tissue/organ building system (MtoBS) with six dispensing heads. In the MtoBS, two heads are for dispensing a synthetic polymer with a heating block and a pneumatic pressure dispenser, and four heads are for dispensing cell-laden hydrogel, with a cooling block and plunger dispenser. They studied the fabrication of cell-printed structures using MtoBS with synthetic polymers (such as PCL, PEG, PLGA and TCP) and natural hydrogels (such as collagen, gelatin, alginate and in-house-developed dECM), and the application of these to the regeneration of osteochondral, ear, liver and heart tissue [15, 19, 24].

References

[1] Crump S S 1991 Fast, precise, safe prototypes with FDM. Intelligent design and manufacturing for prototyping *PED* **50** http://sffsymposium.engr.utexas.edu/Manuscripts/1992/1992-34-Walters.pdf

[2] Crump S S 1992 Apparatus and method for creating three-dimensional objects US Patent No 5,121,329

[3] Crump S S 2002 Direct rapid manufacturing with real production plastics using fused deposition modeling (FDM) *Euro-uRapid Conference (Stratasys, Frankfurt)* http://www.ids-plastics.com/Common/Paper/Paper_82/Direct%20Rapid%20Manufacturing.pdf

[4] Hutmacher D W 2000 Scaffolds in tissue engineering bone and cartilage *Biomaterials* **21** 2529–43

[5] Mironov V, Boland T, Trusk T, Forgacs G and Markwald R R 2003 Organ printing: computer-aided jet-based 3D tissue engineering *Trends Biotechnol.* **21** 157–61

[6] Wang F, Shor L, Darling A, Khalil S, Sun W, Güçeri S and Lau A 2004 Precision extruding deposition and characterization of cellular poly-ϵ-caprolactone tissue scaffolds *Rapid Prototyping J.* **10** 42–9

[7] Mironov V, Visconti R P, Kasyanov V, Forgacs G, Drake C J and Markwald R R 2009 Organ printing: tissue spheroids as building blocks *Biomaterials* **30** 2164–74

[8] Atala A, Bauer S B, Soker S, Yoo J J and Retik A B 2006 Tissue-engineered autologous bladders for patients needing cystoplasty *Lancet* **367** 1241–6

[9] Khalil S and Sun W 2009 Bioprinting endothelial cells with alginate for 3D tissue constructs *J. Biomech. Eng.* **131** 111002

[10] Schuurman W, Khristov V, Pot M W, Van Weeren P, Dhert W J A and Malda J 2011 Bioprinting of hybrid tissue constructs with tailorable mechanical properties *Biofabrication* **3** 021001

[11] Murphy S V and Atala A 2014 3D bioprinting of tissues and organs *Nat. Biotechnol.* **32** 773–5

[12] Jones N 2012 Science in three dimensions: the print revolution *Nature* **487** 22–3

[13] Woodfield T B, Malda J, De Wijn J, Peters F, Riesle J and van Blitterswijk C A 2004 Design of porous scaffolds for cartilage tissue engineering using a three-dimensional fiber-deposition technique *Biomaterials* **25** 4149–61

[14] Kim J Y and Cho D W 2009 Blended PCL/PLGA scaffold fabrication using multi-head deposition system *Microelectron. Eng.* **86** 1447–50

[15] Shim J H, Lee J S, Kim J Y and Cho D W 2012 Bioprinting of a mechanically enhanced three-dimensional dual cell-laden construct for osteochondral tissue engineering using a multi-head tissue/organ building system *J. Micromech. Microeng.* **22** 085014

[16] Zein I, Hutmacher D W, Tan K C and Teoh S H 2002 Fused deposition modeling of novel scaffold architectures for tissue engineering applications *Biomaterials* **23** 1169–85

[17] Hutmacher D W, Schantz T, Zein I, Ng K W, Teoh S H and Tan K C 2001 Mechanical properties and cell cultural response of polycaprolactone scaffolds designed and fabricated via fused deposition modeling *J. Biomed. Mater. Res.* **55** 203–16

[18] Cao T, Ho K H and Teoh S H 2003 Scaffold design and *in vitro* study of osteochondral coculture in a three-dimensional porous polycaprolactone scaffold fabricated by fused deposition modeling *Tissue Eng.* **9** 103–12

[19] Pati F, Jang J, Ha D H, Kim S W, Rhie J W, Shim J H and Cho D W 2014 Printing three-dimensional tissue analogues with decellularized extracellular matrix bioink *Nat. Commun.* **5** 3935

[20] Shim J H, Yoon M C, Jeong C M, Jang J, Jeong S I, Cho D W and Huh J B 2014 Efficacy of rhBMP-2 loaded PCL/PLGA/β-TCP guided bone regeneration membrane fabricated by 3D printing technology for reconstruction of calvaria defects in rabbit *Biomed. Mater.* **9** 065006

[21] Shim J H, Huh J B, Park J Y, Jeon Y C, Kang S S, Kim J Y and Cho D W 2012 Fabrication of blended polycaprolactone/poly (lactic-co-glycolic acid)/β-tricalcium phosphate thin membrane using solid freeform fabrication technology for guided bone regeneration *Tissue Eng.* A **19** 317–28

Shim J H, Kim S E, Park J Y, Kundu J, Kim S W, Kang S S and Cho D W 2014 Three-dimensional printing of rhBMP-2-loaded scaffolds with long-term delivery for enhanced bone regeneration in a rabbit diaphyseal defect *Tissue Eng.* A **20** 1980–92

[22] Wilson W C and Boland T 2003 Cell and organ printing 1: protein and cell printers *Anat. Rec. A* **272** 491–6

[23] Fedorovich N E, Alblas J, de Wijn J R, Hennink W E, Verbout A J and Dhert W J 2007 Hydrogels as extracellular matrices for skeletal tissue engineering: state-of-the-art and novel application in organ printing *Tissue Eng.* **13** 1905–25

[24] Lee J S, Hong J M, Jung J W, Shim J H, Oh J H and Cho D W 2014 3D printing of composite tissue with complex shape applied to ear regeneration *Biofabrication* **6** 024103

Chapter 5

3D printing software

3D printing techniques have been thought of as a new industrial revolution because they have resulted in the flexible and rapid production of highly customized free-form structures without the necessity of spending time and money to modify or organize manufacturing tools and processes. Compared with 'subtractive' manufacturing techniques, such as tooling and cutting, 3D printing is an effective way to create a highly complex 3D structure or provide customized products and devices [1].

Until the 1990s, 3D printing techniques were used in the fabrication of scaffolds for precise and predictable control of micro-architecture and high reproducibility [2]. 3D printing enables the production of implants that are well aligned with an individual patient's unique anatomy and defects [3], but it also enables the design and fabrication of scaffold architecture for effective tissue regeneration [4]. The internal architecture of a scaffold can generally be defined by pore size, interconnectivity, geometry, porosity and internal channel geometry, and these are related to a significant degree to mechanical and biological properties, such as oxygen/nutrient diffusion, cell adhesion rate, tissue formation, mechanical behavior and degradation kinetics [5, 6]. In the 3D printing process, the desired scaffold internal and external architecture can be realized from CAD/ CAM of the scaffold using software subsystems on the basis of hardware components [7].

All 3D printing processes begin with some 3D modeling of the object, generally using a CAD tool. For biomedical applications, medical images, such as CT or MRI, are frequently used to design a specific implant. Subsequently, a CAM system converts 3D CAD models of components into printing path data compatible with a specific 3D printing machine. For tissue engineering applications, porous structure modeling should be performed, either by manual design of a porous solid model using CAD tools [8] or automated generation of a printing path containing a specific pore pattern [9]. Because there are several different ways to design an object, understanding the CAD/CAM system is key for obtaining the desired customized 3D model efficiently.

There are several existing strategies for CAD/CAM, according to the scaffold design, the specific target tissue and the 3D printing apparatus. The main objective of this chapter has been to review efficient and practical methods for the creation of the desired customized constructs using CAD/CAM systems. A brief description of basic CAD/CAM systems is also provided in the following sections.

5.1 Computer-aided scaffold architecture design

5.1.1 Specific external shape design

Construction of a CAD model for a patient-specific target tissue or defect often begins with the acquisition of medical image data from an image processing instrument, such as a 3D scanner or one using CT or MRI [7]. From such images stored in a specific file format (e.g., DICOM), a 3D model for the patient can be extracted/reformed and then exported into the STL file format using medical image processing software, such as MIMICS (Materialise, Inc., MI, USA) [10, 11]. Alternatively, a specific implant model can be designed directly from measurement/ analysis of patient image data using commercial CAD software with a powerful user interface and tools to design a desired structure [12].

5.1.2 Scaffold pore architecture design

As mentioned above, the scaffold's pore architecture plays an important role in regulating its mechanical properties, cell growth, behavior and differentiation, which further affect *in vivo* tissue-forming ability. To meet the requirements for a specific target tissue, the scaffold pore architecture can be designed using two effective strategies.

Following the external shape design, the scaffold pore architecture can be designed using commercial CAD software tools. Complex and irregular pore architectures can be designed without limitations. Boolean operations (e.g., removal or intersection) between the external shape model and the pore architecture model are used to form a scaffold model customized for a patient with a desired pore architecture [13]. However, this is a computationally intensive process that can exceed the capacity of many computers.

In the second strategy, the pore pattern inside an external shape model is defined by a mathematical equation [14]. After a surface or curve-based periodic pore pattern is generated by a specific function, a Boolean operation is carried out to combine the defined pore architecture and the external shape spatially to obtain the final shape of the scaffold. Because this strategy requires lighter computational processing than the first one, it has been preferred for the design of pore architectures with periodic or regular patterns, such as gyroid, diamond [15] and 0°/90° lay-down [16] architectures.

5.2 Automated printing path generation

5.2.1 Slicing and printing path generation algorithms

3D printing (or additive manufacturing) is based on stacking up thin cross-sections of an object, layer by layer. Thus, each of these layers should be calculated from a CAD model by a horizontal slicing algorithm [17]. These sliced layers would be a closed loop, composed of many tiny lines, obtained from an intersection calculation between a 3D model of polygons from the STL file data and specific horizontal planes.

Printing paths are then defined by intersection calculations. Specifically oriented lines at regular intervals are placed on each layer. Each line can be defined by a constraint equation (e.g., $(x = x_i, z = z_{2n})$ or $(y = y_j, z = z_{2m+1})$). Their end points are decided by intersection calculations between the lines and the closed loops, generated by a slicing algorithm. The lines should also be arranged to minimize the total printing time.

One major parameter in printing path generation is line distance. This parameter value should be larger than the line width to avoid a collision between printed lines. If the expected line width and the line distance are the same, the printed structure will have the same shape as the CAD model. However, as shown in figure 5.1, a lay-down pore pattern can be obtained by using a line distance that is larger than the line width. Then, the pore size is defined by the difference between the line distance and the line width.

5.2.2 Printing path generation for 3D printing of cell-laden hydrogel

For the creation of a 3D biofunctional tissue/organ construct, the arrangement of various kinds of cells at desired locations should be considered. 3D printing of cell-laden hydrogel constructs has been proposed as an innovative way to bio-mimic the complex structure of cells and matrices, such as heterogeneous tissues [18]. In the

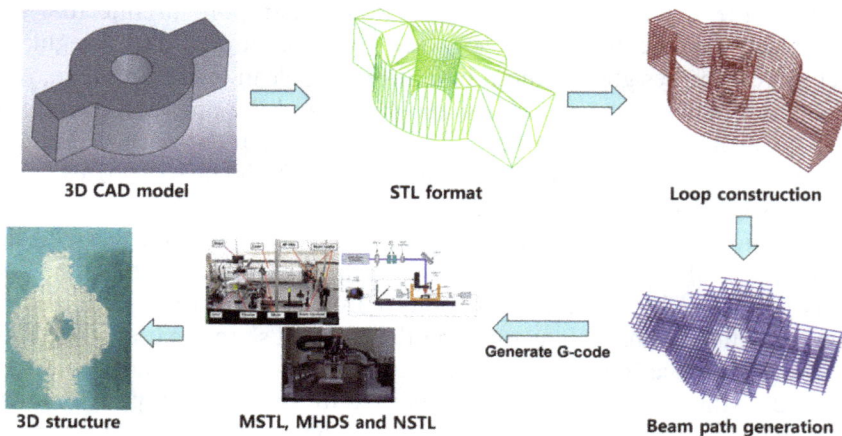

Figure 5.1. Printing path generation process for single material printing.

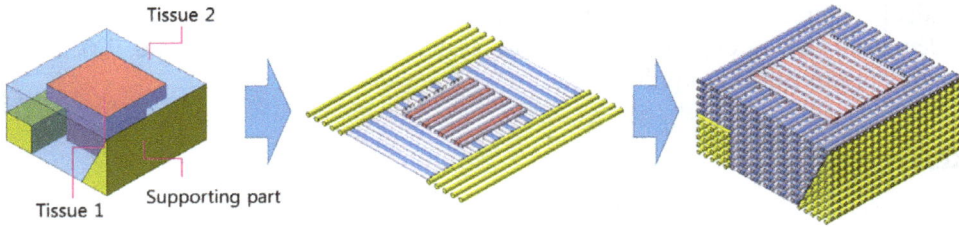

Figure 5.2. Multi-material printing of a 3D cell-laden structure with a frame component.

printing of hydrogel polymers containing cells, a printed layer of pre-gel solution in the liquid state should be instantly crosslinked to retain structural integrity, which is needed to stack the next 2D cell-patterned layer [19]. The printing path for multiple cell-laden hydrogels is divided spatially according to the constitution of the specific target tissue.

However, a 3D structure composed only of hydrogels is likely to be too weak to maintain its own shape against gravity and external forces. It has been suggested that printing a synthetic polymer as a supporting frame might avoid these problems when fabricating cell-laden structures [20]. The cell/hydrogel printing path is simply made in the gap between synthetic polymer printing paths, as shown in figure 5.2. The synthetic polymer, as a frame, prevents the cell/hydrogel path from collapsing until after the printing process is complete. With this printing method, multiple cell types can be delivered to locations that have been pre-determined with a CAD/CAM system.

5.2.3 Image-based 3D printing path generation

For rapid production, many reports have suggested an image-based 3D printing system, which prints 2D patterns on a layer from one or a few image projections [21]. A digitalized image is generally composed of pixels, a square-shaped point having color information. Thus, the printing path for image-based printing can be composed of white and black pixels located on the inside and outside of a layer, respectively [17].

5.3 Conclusion

Today, 3D printing has been recognized as a revolutionary means for realizing 'made-to-order' tissues/organs in the field of medicine. To spread 3D printing techniques widely to clinicians, or to create hybrid scaffolds for the reconstruction of complex organs, optimization strategies for easy and intuitive custom designs using CAD/CAM are needed. For simple and versatile CAD/CAM strategies, we suggest the use of commercially available CAD software, and the development of automated CAM adapted for a specific machine. The customized external shape is designed with the 3D modeling tools of the CAD software, usually with CT or MRI data. Subsequently, the developed CAM system automatically generates the desired internal architecture for the 3D model. However, the CAD/CAM systems developed recently to create complex tissue structures are not yet familiar to most medical/bio-engineering researchers. To provide more convenient and easy-to-understand user interfaces, CAD/CAM systems need to be explored further for various organ printing applications.

References

[1] Berman B 2012 3D printing: the new industrial revolution *Bus. Horiz.* **55** 155–62
[2] Hutmacher D W 2000 Scaffolds in tissue engineering bone and cartilage *Biomaterials* **21** 2529–43
[3] Rengier F *et al* 2010 3D printing based on imaging data: review of medical applications *Int. J. Comput. Ass. Rad.* **5** 335–41
[4] Seol Y J, Kang T Y and Cho D W 2012 Solid freeform fabrication technology applied to tissue engineering with various biomaterials *Soft Matter.* **8** 1730–5
[5] Hutmacher D W, Sittinger M and Risbud M V 2004 Scaffold-based tissue engineering: rationale for computer-aided design and solid free-form fabrication systems *Trends Biotechnol.* **22** 354–62
[6] Jung J W *et al* 2013 Evaluation of the effective diffusivity of a freeform fabricated scaffold using computational simulation *J. Biomech. Eng.* **2013** 135
[7] Sun W, Starly B, Nam J and Darling A 2005 Bio-CAD modeling and its applications in computer-aided tissue engineering *Computer-Aided Design.* **37** 1097–114
[8] Cheah C M, Chua C K, Leong K F, Cheong C H and Naing M W 2004 Automatic algorithm for generating complex polyhedral scaffold structures for tissue engineering *Tissue Eng.* **10** 595–610
[9] Lee J S, Cha H D, Shim J H, Jung J W, Kim J Y and Cho D W 2012 Effect of pore architecture and stacking direction on mechanical properties of solid freeform fabrication-based scaffold for bone tissue engineering *J. Biomed. Mater. Res.* A **100** 1846–53
[10] Moerenhout BAMML Gelaude F, Swennen G R J, Casselman J W, Van der Sloten J and Mommaerts M Y 2009 Accuracy and repeatability of cone-beam computed tomography (CBCT) measurements used in the determination of facial indices in the laboratory setup *J. Cranio. Maxill. Surg.* **37** 18–23
[11] Singare S *et al* 2009 Rapid prototyping assisted surgery planning and custom implant design *Rapid Prototyping J.* **15** 19–23
[12] Kim T H *et al* 2014 *In vitro* and *in vivo* evaluation of bone formation using solid freeform fabrication-based bone morphogenic protein-2 releasing PCL/PLGA scaffolds *Biomed. Mater.* **2014** 9
[13] Jung J W, Park J H, Hong J M, Kang H W and Cho D W 2014 Octahedron pore architecture to enhance flexibility of nasal implant-shaped scaffold for rhinoplasty *Int. J. Precis. Eng. Man.* **15** 2611–6
[14] Yoo D 2013 New paradigms in hierarchical porous scaffold design for tissue engineering *Mat. Sci. Eng.* C **33** 1759–72
[15] Melchels F P W, Bertoldi K, Gabbrielli R, Velders A H, Feijen J and Grijpma D W 2010 Mathematically defined tissue engineering scaffold architectures prepared by stereolithography *Biomaterials* **31** 6909–16
[16] Zein I, Hutmacher D W, Tan K C and Teoh S H 2002 Fused deposition modeling of novel scaffold architectures for tissue engineering applications *Biomaterials* **23** 1169–85

[17] Jung J W, Kang H W, Kang T Y, Park J H, Park J and Cho D W 2012 Projection image-generation algorithm for fabrication of a complex structure using projection-based micro-stereolithography *Int. J. Precis. Eng. Man.* **13** 445–9

[18] Khalil S and Sun W 2007 Biopolymer deposition for freeform fabrication of hydrogel tissue constructs *Mater. Sci. Eng.* C **27** 469–78

[19] Moon S *et al* 2010 Layer by layer three-dimensional tissue epitaxy by cell-laden hydrogel droplets *Tissue Eng.* C **16** 157–66

[20] Shim J H, Kim J Y, Park M, Park J and Cho D W 2011 Development of a hybrid scaffold with synthetic biomaterials and hydrogel using solid freeform fabrication technology *Biofabrication* **2011** 3

[21] Kang H W, Seol Y J and Cho D W 2009 Development of an indirect solid freeform fabrication process based on microstereolithography for 3D porous scaffolds *J. Micromech. Microeng.* **2009** 19

Chapter 6

Biomaterials for organ printing

Polymers are used widely as biomaterials for the bioprinting of tissue engineering scaffolds and cell-laden constructs [1] because they offer distinct advantages in terms of cytocompatibility, versatility of chemistry and biological properties that are important in tissue engineering and regenerative medicine [2]. Biopolymers are classified into several types with respect to their structural, chemical and biological characteristics. Based on their origin, they are classified as natural, synthetic and semi-synthetic. Two types of biopolymers are used most often for bioprinting. The first are curable polymers, which result in mechanically robust and durable materials. They are used to produce scaffolds or to print supporting frameworks along with other materials to fabricate constructs. They typically require high temperatures or toxic solvents for their processing and printing, and are thus not suitable for printing together with cells. Post-printing cell seeding is generally used for scaffolds prepared with these polymers, avoiding conditions harmful to the cells. Hydrogels are the second type and these can be of synthetic or natural origin. Their composition and structure are similar to those of natural tissue and they can generate tissue mimics [3]. Hydrogels are composed mainly of water (up to >99%), with the remainder of their make-up being a hydrophilic polymer network that confines the water within its boundaries [4]. Hydrogels are also popular for encapsulating living cells because they can be processed using relatively mild conditions and aqueous chemistries [5]. However, hydrogels do not possess the same levels of mechanical properties as curable support polymers and they are mostly used for engineering soft tissues. There are many important characteristics, such as mechanical properties, melting points, available chemistries for crosslinking, gelation kinetics and potential functionalization that determine the usefulness of polymers in bioprinting. In the next few chapters, we provide an overview of some of the traditional materials used in bioprinting and several other materials that have been developed for the purpose of bioprinting.

doi:10.1088/978-1-6817-4079-9ch6

References

[1] Melchels F P W, Domingos M A N, Klein T J, Malda J, Bartolo P J and Hutmacher D W 2012 Additive manufacturing of tissues and organs *Prog. Polym. Sci.* **37** 1079–104

[2] Lee H B, Khang G and Lee J H 2003 Polymeric biomaterials *Biomaterials: Principles and Applications* ed J B Park and J D Bronzino (Boca Raton, FL: CRC) www.crcpress.com/Biomaterials-Principles-and-Applications/Park-Bronzino/9780849314919

[3] Slaughter B V, Khurshid S S, Fisher O Z, Khademhosseini A and Peppas N A 2009 Hydrogels in regenerative medicine *Adv. Mater.* **21** 3307–29

[4] Calvert P 2009 Hydrogels for soft machines *Adv. Mater.* **21** 743–56

[5] Drury J L and Mooney D J 2003 Hydrogels for tissue engineering: scaffold design variables and applications *Biomaterials* **24** 4337–51

IOP Concise Physics

Organ Printing

Dong-Woo Cho, Jung-Seob Lee, Falguni Pati, Jin Woo Jung, Jinah Jang and Jeong Hun Park

Chapter 7

Natural, synthetic and semi-synthetic polymers

7.1 Natural polymers

Natural polymers are a class of biomaterials that are typically isolated from natural sources and are used widely in biomedical applications. Natural polymers have better interactions with cells due to their biological recognition, which is vital for encouraging cellular interaction and superior biological performance [1]. Biocompatible natural polymers show minimal inflammatory responses at an implantation site [2]. Biopolymers can be of human or non-human origin; human-derived biopolymers, such as collagen and fibrin, have the best biological compatibility and degrade via proteolytic pathways (proteolysis by specific enzymes), while non-human-derived biopolymers, such as alginates and chitosan, have moderate biological compatibility and degrade mainly via hydrolytic pathways [2]. Nevertheless, many human-derived biopolymers are preferentially produced from non-human sources because they are more readily and economically available. For example, hyaluronan, a glycosaminoglycan found in humans, is much more efficiently produced from bacteria [3]. Traditionally, plant-derived biopolymers (e.g., alginates, agarose, cellulose) have been inexpensive, abundantly available and widely used in cell culture [4]. However, as they are of non-human origin, they are moderately biocompatible and degrade via non-proteolytic mechanisms within the body [3]. The major drawbacks of naturally occurring biopolymers are significant variations in molecular weight and structure from batch to batch, and the potential risk of pathogen transfer from the originating organism [5]. Natural polymers can be classified based on their composition as proteins (collagen, gelatin, silk and fibrin), polysaccharides (alginates, agarose and chitosan) and glycosaminoglycans.

Agarose is a polysaccharide consisting of a galactose-based backbone, and is extracted from seaweed. It is commonly used as a cell culture medium component. It is an attractive biopolymer because its stiffness can be altered, which means that the mechanical properties of the scaffold can be tuned. It is generally used for cell

doi:10.1088/978-1-6817-4079-9ch7

encapsulation and provides a uniform cellular distribution throughout the scaffolds. The degradation rate of this polymer is relatively slow, similar to that of alginates. Agarose exhibits a temperature-sensitive solubility in water and has found widespread applications in tissue engineering [6]. It has been printed as follows: the polymer solution is held in the printer reservoir at 60–80 °C and printed into a cool bath below the gel transition temperature [7, 8]. However, a limitation of this approach is that the structures tend to be weak and may need to be reinforced using other polymers or with a post-print crosslinking step. In some instances, the initial and final temperature of the polymer solution/gel may also prevent it from being able to include cells during printing because temperatures beyond physiological ranges could damage the cells [9].

Alginates are polysaccharides derived from algae and consist of two repeating monosaccharides, such as L-guluronic acid and D-mannuronic acids. Alginate is a polyanionic polymer that forms a firm ionotropic hydrogel upon the addition of calcium ions [10, 11]. This polymer is particularly attractive for the encapsulation of chondrocytes because their phenotype in culture can be maintained, which allows cells that have been cultured in monolayers to be re-differentiated [12]. Calcium alginate scaffolds are not degraded by hydrolytic cleavage; instead, they are degraded by chelating agents, such as EDTA, and certain enzymes [13]. Alginate has been printed into a calcium solution to produce microspheres as well as more complex structures [14, 15]. The main advantage of the reactive printing of ionotropic polymers like alginate is the very rapid gelation (~1 s), which ensures the stability of the printed structure soon after fabrication [16].

Chitosan is a deacetylated derivative of chitin, which is a naturally occurring polysaccharide extracted from crab shells, shrimps and fungi [17]. Chitin and chitosan are semicrystalline polymers and possess a high degree of biocompatibility [18]. Chitosan is a polycationic polymer that contains glucosamine residues, which are positively charged above its isoelectric point. Thus, chitosan forms a firm ionotropic hydrogel with phosphate ions [19]. A temperature-sensitive carrier material can also be prepared from chitosan and can be used as an injectable. It forms gels at body temperature and has the ability to deliver and interact with growth factors and adhesion proteins [20]. The degradation of chitosan can be controlled by its residual acetyl content and its degradation rate is relatively fast *in vivo*. Chitosan hydrogels were used in bioprinting to fabricate vessel-like tubular microfluidic channels [21].

Collagen, the major structural protein of humans, is widely used for biomedical and pharmaceutical applications [22]. There are more than 22 types of collagen found in the human body, among which types I–IV are the most widely studied [22]. The isolation and purification processes for collagens are well established, particularly for collagen type I, so using collagen materials for surface coatings and the manufacture of gels for cell culture environments has become common. Collagen provides cellular recognition due to the presence of arginine–glycine–aspartic acid (RGD) amino acid sequences, which form motifs that allow cells to adhere and proliferate via integrin–RGD binding [23]. The main advantage of collagens is that they degrade at a rate that more closely matches that of cellular growth because cells are programmed to produce the proteolytic enzyme collagenase to make room for growth [24]. Collagen has been used extensively in many forms, such as fibers, gels, solutions, filamentous materials, tubular materials (membranes and sponges) and composite matrices, for tissue engineering applications [25]. Composite scaffolds of calcium phosphate and collagen were prepared by 3D printing and used for bone regeneration, with new bone growth incorporating the degrading scaffold materials [26]. Collagen was also used to fabricate structures by 3D bioprinting to control BMP-2 and VEGF delivery spatially and temporally to promote large-volume bone regeneration [27].

Fibrin plays an important role in wound healing and prevents blood loss during injury. Fibrin glue is formed from the mixture of fibrinogen and thrombin that allows solidification. It is often used as a carrier for cells and in conjunction with other scaffold materials. Fibrin gels can be degraded by hydrolytic or enzymatic processes. Fibrin gels have applications as tissue engineering scaffold matrices, especially for the repair of cartilage tissue [28]. 3D bioprinting of collagen and VEGF-releasing fibrin gel scaffolds were used for neural stem cell culture and the results demonstrated that bioprinting of VEGF-containing fibrin gel supported sustained release of the VEGF in the collagen scaffold [29]. *In situ* printing of fibrin-collagen gels with amniotic fluid-derived stem cells induced increased wound closure rates as well as increased vascularization of the regenerating tissue [30].

Gelatin is derived from collagen and exhibits minimal immune responses. Gelatin contains many glycine, proline and 4-hydroxyproline residues. Gelatin is a denatured protein obtained by the acid and alkaline processing of collagen. Due to its easy processability and its gelation properties, gelatin is popular for 3D bioprinting [31]. Gelatin matrices incorporating cell adhesion factors, such as vitronectin, fibronectin, and RGD peptides, enhance cell proliferation. Gelatin matrices have been used to treat and regenerate tissues, such as bone, cartilage, adipose tissue and skin [32]. Gelatin is chemically modified to incorporate methacrylate groups to enable cross-linking by UV light [33]. Gelatin has been used in several ways in bioprinting, including employment of its temperature-sensitivity to facilitate dispensing and printing. Heterogeneous aortic valve conduits were prepared by 3D bioprinting with alginate/gelatin hydrogels, with direct encapsulation of smooth muscle cells (SMC) in the valve root and aortic valve leaflet interstitial cells (VIC) in the leaflets [15]. Photolabile cell-laden methacrylated gelatin (GelMA) hydrogel was bioprinted with varying cell densities. It was reported that encapsulated HepG2 cells showed cell viability for at least eight days following the bioprinting process [31].

Hyaluronan, or hyaluronic acid (HA), is an anionic polysaccharide that is used as a cell carrier for regenerative medicine. HA is found abundantly within cartilaginous ECMs. It can be crosslinked to form scaffolds and seeded with chondrocytes and stem cells to induce chondrogenesis and osteogenesis on the scaffold [34]. It can be used as an injectable to fill irregularly shaped defects with minimal invasion. Moreover, HA is not antigenic and elicits no inflammatory or foreign body reaction [35]. Traditionally, HA has been used in clinics in applications such as therapy for damaged joints and arthritis [36]. However, chemically modified HA has found widespread applicability due to the formation of a robust biomaterial that can be crosslinked into a hydrogel or loaded with other bioactive factors [37]. HA hydrogels are commonly modified with photocrosslinkable methacrylate groups that can undergo free radical polymerization when exposed to UV light, resulting in soft hydrogels. Photocrosslinkable methacrylated HA (MA-HA) hydrogels have found many applications, such as in cutaneous and corneal wound healing [38] and for bioprinting of prototype vessel structures [39].

Silks are protein polymers spun by insects, such as silkworms and spiders. Silkworm silk consists of two types of proteins: fibroin and sericin. Silk fibroin has remarkable mechanical properties, biocompatibility and controlled degradation rates, and it can be modified chemically to alter its surface properties [40]. The conformational and eventual morphology of the silk fibroin chain are critical in the improved mechanical properties [41]. Silk fibroin matrices are fabricated from regenerated silk solutions to form films, gels and 3D scaffolds. Silk fibroin scaffolds have been shown to support cell adhesion, proliferation and differentiation *in vitro* and *in vivo* to engineer a range of tissues, such as bone, cartilage, tendons and skin [42]. Silk fibroin scaffolds were prepared by direct-write assembly for tissue engineering applications [43]. Regenerated silk solution was used to print structures in a methanol bath because silk is insoluble in methanol. However, cells cannot be used in the printing process and cells are generally seeded after printing the structure. Silk–gelatin microperiodic 3D scaffolds were prepared by 3D bioprinting and supported the redifferentiation of chondrocytes; they have potential for use in cartilage tissue engineering applications [44]. Our group has prepared a bioprintable, cell-laden silk fibroin–gelatin hydrogel that supports the multilineage differentiation of stem cells. Cell are loaded in the hydrogel and used to fabricate 3D tissue constructs with high cell viability and tissue-forming capacity [45].

7.2 Synthetic and semi-synthetic polymers

Biodegradable synthetic polymers are man-made materials that have found many applications in the biomedical field because of their tailorable properties. They are particularly suitable for tissue engineering applications, because 3D structures of various shapes and sizes can be fabricated from synthetic polymers [46]. The physicochemical and mechanical properties of the synthetic polymers can be tuned to match certain biological tissues for better outcomes [47]. Synthetic polymers are generally degraded by simple hydrolysis, and the degradation rate can be tailored based on their molecular weight and composition. These polymers can be produced consistently and cheaply in large quantities and, above all, they are easy to modify to produce hydrogels with desirable properties. Most importantly, the risk of pathogens being present in a synthetic polymer is negligible. However, care must be taken to remove/exclude all traces of toxic unpolymerized/uncrosslinked reagents left in the hydrogel prior to use [48]. Most synthetic polymers are not biocompatible and have limited biodegradability and poor cellular adhesion; however, many of these shortcomings have been addressed to some extent with appropriate processing and modification strategies [49, 50]. The most extensively used synthetic polymers are poly(glycolic acid) (PGA), poly(lactic acid) (PLA) and their co-polymers poly (lactide-co-glycolide), polycaprolactone (PCL), poly(propylene fumarate), poly-ethylene glycol (PEG) and polyurethane.

Poly(α-hydroxy ester)s are used as biodegradable scaffold materials due to their good biocompatibility, controlled biodegradability and relatively good processability. These polymers degrade by non-specific hydrolytic scission of the ester bond, producing glycolic acid. The degradation rate of PLA is slower than that for PGA due to its hydrophobic characteristics, and PLA–PGA co-polymers show no linear relationship with the ratio of PLA to PGA with regard to their degradation. PGA, PLA and PLGA scaffolds have been used in the regeneration of many tissues, including skin, cartilage, blood vessels, nerves and liver [51]. FDM has been used to fabricate scaffolds with highly interconnecting and controllable pore structures, mainly for bone tissue engineering applications using PLA and PLGA and their composites with ceramics [52]. An HA/PLGA conjugate scaffold was also 3D-printed with an intact BMP-2/PEG complex for bone tissue regeneration [53]. Osteochondral composite constructs were 3D-printed in which the upper region was composed of D, L-PLGA/L-PLA with 90% porosity for cartilage regeneration, and the lower region was a LPLGA/TCP composite to maximize bone ingrowth [54].

Polycaprolactone is a thermoplastic with a low melting point (59–64 °C), a form that blends readily with other polymers and is commonly used in bioprinting as a scaffolding component. PCL is non-toxic, biocompatible and degrades through hydrolytic scission, with resistance to rapid hydrolysis. The degradation rate of PCL is rather slow; its complete degradation can take as long as 24 months. Thus, it is co-polymerized with other materials to generate the desired degradation properties. Due to its excellent biocompatibility, PCL has also been investigated extensively as a scaffold for tissue engineering [55]. PCL has been used extensively for making scaffolds using dispensing-based 3D printing techniques. A blend of PCL and PLGA has been used to make 3D scaffolds using a multi-head deposition system [56], where molten PCL is extruded from a micronozzle and deposited on a cool stage. PCL-alginate-chondrocyte bioprinted scaffolds were printed by a dispensing-based 3D printer and used for cartilage tissue engineering applications [57]. They have shown that the printing of molten PCL does not cause much cell death; the printing process is cytocompatible. PCL has also been used for 3D printing of composite tissues with complex shapes, targeted to ear regeneration, along with MSCs encapsulated in an alginate hydrogel [58]. A blend of polycaprolactone/poly(lactic-co-glycolic acid)/b-tricalcium phosphate has been used to make a thin membrane using 3D printing technology for guided bone regeneration [59]. Bioprinting of a mechanically enhanced 3D dual cell-laden construct using PCL as a supporting framework, and alginate as a cell carrier, was prepared using a multi-head tissue/organ building system for osteochondral tissue engineering [60]. Solid laser sintering (SLS) was used to fabricate scaffolds from PCL to successfully construct prototypes of mini-pig mandibular condyle scaffolds [52]. Biocomposite blends of PCL with different percentages of HA were subjected to an evaluation of their suitability for fabrication via SLS, which can also replicate the desired anatomy precisely [61].

Poly(propylene fumarate) and its copolymers are biodegradable unsaturated linear polyesters. These polymers generally degrade via hydrolytic chain scission, similar to that of poly(α-hydroxy esters). The degradation behavior and mechanical strength of these polymers can be controlled by crosslinking of vinyl monomers with the unsaturated double bonds. The physical properties of poly(propylene fumarate) are enhanced by making composites with calcium-phosphate-based degradable bioceramics and they have been used as a bone substitute material [62]. Poly(propylene fumarate) scaffolds with bone morphogenetic protein 2 (BMP-2) were fabricated by 3D printing technology [63]. These scaffolds showed promising results for bone tissue regeneration. SLS was used to fabricate micropatterned scaffolds for bone tissue engineering applications using a blend of diethyl fumarate and poly(propylene fumarate) [53].

Polyethylene glycol (PEG) is a hydrophilic synthetic polymer that restricts and controls the attachment of cells and proteins on scaffolds. PEG, being hydrophilic, prevents the adherence of other proteins to the scaffold surface and thereby minimizes any adverse immune response. Using PEG in a co-polymer, researchers can control the cell attachment characteristics of the scaffolds and enhance the biocompatibility of the co-polymer [64]. PEG can be modified to attach acrylate groups to enable UV crosslinkability, and acrylated PEG (PEG-diacrylate or PEGDA) has been used extensively for fabricating cell-laden micropatterned structures [52]. PEG was also used to make sacrificial components by 3D printing and this method was used for generating human-ear-shaped constructs with complex shapes [58].

Polyvinyl alcohol (PVA) is a biocompatible synthetic polymer that can swell to hold a large amount of water. It possesses reactive pendant alcohol groups and can be modified easily using physical or chemical crosslinking. It can be transformed easily to produce hydrogels, the properties of which can be tailored. PVA hydrogels have applications in cartilage regeneration, breast augmentation and diaphragm replacement [65]. However, a limitation of using PVA scaffolds is their incomplete degradation [66]. PVA has been used to make replicas of a patient's liver by 3D printing to determine how to best carve a donor liver with minimal tissue loss to fit a recipient's abdominal cavity [67]. PVA was used because it has a water content and texture similar to living tissues, allowing a more realistic penetration by surgical blades.

Polyurethane is one of the most widely used polymeric biomaterials in the biomedical field due to its unique physical properties, such as its durability, elasticity, elastomer-like characteristics, fatigue resistance, compliance and tolerance. The degradation of polyurethane takes place by hydrolysis, oxidation and thermal and enzymatic means. Foreign-body reactions are not induced and antibody formation has not been observed when this polymer is used for *in vivo* applications [68]. Biodegradable elastomeric polyurethane scaffolds have been fabricated by an inkjet-based 3D printing technique [69].

References

[1] Kim M S, Kim J H, Min B H, Chun H J, Han D K and Lee H B 2011 Polymeric scaffolds for regenerative medicine *Polym. Rev.* **51** 23–52

[2] Pariente J-L, Kim B-S and Atala A 2001 *In vitro* biocompatibility assessment of naturally derived and synthetic biomaterials using normal human urothelial cells *J. Biomed. Mater. Res.* **55** 33–9

[3] Rinaudo M 2008 Main properties and current applications of some polysaccharides as biomaterials *Polym. Int.* **57** 397–430

[4] Khademhosseini A and Langer R 2007 Microengineered hydrogels for tissue engineering *Biomaterials* **28** 5087–92

[5] Collier J H and Segura T 2011 Evolving the use of peptides as components of biomaterials *Biomaterials* **32** 4198–204

[6] Finger A R, Sargent C Y, Dulaney K O, Bernacki S H and Loboa E G 2007 Differential effects on messenger ribonucleic acid expression by bone marrow-derived human mesenchymal stem cells seeded in agarose constructs due to ramped and steady applications of cyclic hydrostatic pressure *Tissue Eng.* **13** 1151–8

[7] Landers R, Pfister A, Hübner U, John H, Schmelzeisen R and Mülhaupt R 2002 Fabrication of soft tissue engineering scaffolds by means of rapid prototyping techniques *J. Mater. Sci.* **37** 3107–16

[8] Landers R, Hübner U, Schmelzeisen R and Mülhaupt R 2002 Rapid prototyping of scaffolds derived from thermoreversible hydrogels and tailored for applications in tissue engineering *Biomaterials* **23** 4437–47

[9] Shoichet M S, Li R H, White M L and Winn S R 1996 Stability of hydrogels used in cell encapsulation: an *in vitro* comparison of alginate and agarose *Biotechnol. Bioeng.* **50** 374–81

[10] Hatefi A and Amsden B 2002 Biodegradable injectable *in situ* forming drug delivery systems *J. Control. Rel.* **80** 9–28

[11] Park S, Lee S and Kim W 2011 Fabrication of hydrogel scaffolds using rapid prototyping for soft tissue engineering *Macromol. Res.* **19** 694–8

[12] Williams G M, Klein T J and Sah R L 2005 Cell density alters matrix accumulation in two distinct fractions and the mechanical integrity of alginate-chondrocyte constructs *Acta Biomaterialia* **1** 625–33

[13] Ashton R S, Banerjee A, Punyani S, Schaffer D V and Kane R S 2007 Scaffolds based on degradable alginate hydrogels and poly(lactide-co-glycolide) microspheres for stem cell culture *Biomaterials* **28** 5518–25

[14] Boland T, Xu T, Damon B and Cui X 2006 Application of inkjet printing to tissue engineering *Biotechnology* J **1** 910–7

[15] Duan B, Hockaday L A, Kang K H and Butcher J T 2013 3D Bioprinting of heterogeneous aortic valve conduits with alginate/gelatin hydrogels *J. Biomed. Mater. Res.* A **101** 1255–64

[16] Fedorovich N E, Alblas J, de Wijn J R, Hennink W E, Verbout A J and Dhert W J 2007 Hydrogels as extracellular matrices for skeletal tissue engineering: state-of-the-art and novel application in organ printing *Tissue Eng.* **13** 1905–25

[17] Goosen M F A 1997 *Applications of Chitin and Chitosan* (Lancaster: Technomic)

[18] Nolan K, Millet Y, Ricordi C and Stabler C L 2008 Tissue engineering and biomaterials in regenerative medicine *Cell Transplant.* **17** 241–3

[19] Hennink W E and van Nostrum C F 2002 Novel crosslinking methods to design hydrogels *Adv. Drug Del. Rev.* **54** 13–36

[20] Shi C M, Zhu Y, Ran X Z, Wang M, Su Y P and Cheng T M 2006 Therapeutic potential of chitosan and its derivatives in regenerative medicine *J. Surg. Res.* **133** 185–92

[21] Zhang Y, Yu Y and Ozbolat I T 2013 Direct bioprinting of vessel-like tubular microfluidic channels *J. Nanotechnol. Eng. Med.* **4** 0210011

[22] Glowacki J and Mizuno S 2008 Collagen scaffolds for tissue engineering *Biopolymers* **89** 338–44

[23] Kadler K E, Baldock C, Bella J and Boot-Handford R P 2007 Collagens at a glance *J. Cell Sci.* **120** 1955–8

[24] Barker T H 2011 The role of ECM proteins and protein fragments in guiding cell behavior in regenerative medicine *Biomaterials* **32** 4211–4

[25] Nair L S and Laurencin C T 2007 Biodegradable polymers as biomaterials *Prog. Polym. Sci.* **2007** 32

[26] Inzana J A *et al* 2014 3D printing of composite calcium phosphate and collagen scaffolds for bone regeneration *Biomaterials* **35** 4026–34

[27] Park J Y *et al* 2015 3D printing technology to control BMP-2 and VEGF delivery spatially and temporally to promote large-volume bone regeneration *J. Mater. Chem.* B **3** 5415–25

[28] Zhao H, Ma L, Zhou J, Mao Z, Gao C and Shen J 2008 Fabrication and physical and biological properties of fibrin gel derived from human plasma *Biomed. Mater. Eng.* **3** 15001

[29] Lee Y-B *et al* 2010 Bio-printing of collagen and VEGF-releasing fibrin gel scaffolds for neural stem cell culture *Exp. Neurol.* **223** 645–52

[30] Skardal A *et al* 2012 Bioprinted amniotic fluid-derived stem cells accelerate healing of large skin wounds *Stem Cells Translational Med.* **1** 792–802

[31] Bertassoni L E *et al* 2014 Direct-write bioprinting of cell-laden methacrylated gelatin hydrogels *Biofabrication* **6** 024105

[32] Lien S-M, Ko L-Y and Huang T-J 2009 Effect of pore size on ECM secretion and cell growth in gelatin scaffold for articular cartilage tissue engineering *Acta Biomaterialia* **6** 670–9

[33] Schuurman W *et al* 2013 Gelatin-methacrylamide hydrogels as potential biomaterials for fabrication of tissue-engineered cartilage constructs *Macromol. Biosci.* **13** 551–61

[34] Kogan G, Soltes L, Stern R and Gemeiner P 2007 Hyaluronic acid: a natural biopolymer with a broad range of biomedical and industrial applications *Biotechnol. Lett.* **29** 17–25

[35] Gerecht S, Burdick J, Ferreira L, Townsend S, Langer R and Vunjak-Novakovic G 2007 Hyaluronic acid hydrogel for controlled self-renewal and differentiation of human embryonic stem cells *Proc. Natl Acad. Sci. USA* **104** 11298

[36] Collins M N and Birkinshaw C 2013 Hyaluronic acid based scaffolds for tissue engineering—a review *Carbohyd. Polym.* **92** 1262–79

[37] Prestwich G D and Kuo J W 2008 Chemically-modified HA for therapy and regenerative medicine *Curr. Pharm. Biotechnol.* **9** 242–5

[38] Miki D *et al* 2002 A photopolymerized sealant for corneal lacerations *Cornea* **2002** 21

[39] Skardal A, Zhang J, McCoard L, Xu X, Oottamasathien S and Prestwich G D 2010 Photocrosslinkable hyaluronan–gelatin hydrogels for two-step bioprinting *Tissue Eng.* A **16** 2675–85

[40] Wray L S *et al* 2011 Effect of processing on silk-based biomaterials: reproducibility and biocompatibility *J. Biomed. Mater. Res.* B **99** 89–101

[41] Kim U J, Park J, Li C, Jin H J, Valluzzi R and Kaplan D L 2004 Structure and properties of silk hydrogels *Biomacromolecules* **5** 786–92

[42] Rockwood D N, Preda R C, Yucei T, Wang X, Lovett M L and Kaplan D L 2011 Materials fabrication from *Bombyx mori* silk fibroin *Nat. Protocols* **6** 1612–31

[43] Ghosh S, Parker S T, Wang X, Kaplan D L and Lewis J A 2008 Direct-write assembly of microperiodic silk fibroin scaffolds for tissue engineering applications *Adv. Funct. Mater.* **18** 1883–9

[44] Das S *et al* 2013 Enhanced Redifferentiation of chondrocytes on microperiodic silk/gelatin scaffolds: toward tailor-made tissue engineering *Biomacromolecules* **14** 311–21

[45] Das S *et al* 2014 Bioprintable, cell-laden silk fibroin-gelatin hydrogel supporting multilineage differentiation of stem cells for fabrication of 3D tissue constructs *Acta Biomaterialia* **11** 233–46

[46] Kim B-S and Mooney D J 1998 Development of biocompatible synthetic extracellular matrices for tissue engineering *Trends Biotechnol.* **16** 224–30

[47] Lutolf M P and Hubbell J A 2005 Synthetic biomaterials as instructive extracellular microenvironments for morphogenesis in tissue engineering *Nat. Biotech.* **23** 47–55

[48] Sung H-W, Huang R-N, Huang L L H and Tsai C-C 1999 *In vitro* evaluation of cytotoxicity of a naturally occurring cross-linking reagent for biological tissue fixation *J. Biomater. Sci. Polym. Ed.* **10** 63–78

[49] Baldwin A D and Kiick K L 2010 Polysaccharide-modified synthetic polymeric biomaterials *Peptide Sci.* **94** 128–40

[50] Tessmar J K and Göpferich A M Customized PEG-derived copolymers for tissue-engineering applications *Macromol. Biosci.* **7** 23–39

[51] Sun H, Qu Z, Guo Y, Zang G and Yang B 2007 *In vitro* and *in vivo* effects of rat kidney vascular endothelial cells on osteogenesis of rat bone marrow mesenchymal stem cells growing on polylactide-glycolic acid (PLGA) scaffolds *Biomed. Eng. Online* **6** 41

[52] Zorlutuna P, Jeong J H, Kong H and Bashir R 2011 Tissue engineering: stereolithography-based hydrogel microenvironments to examine cellular interactions *Adv. Funct. Mater.* **21** 3597

[53] Seol Y-J, Kang T-Y and Cho D-W 2012 Solid freeform fabrication technology applied to tissue engineering with various biomaterials *Soft Matter.* **8** 1730–5

[54] Sherwood J K *et al* 2002 A three-dimensional osteochondral composite scaffold for articular cartilage repair *Biomaterials* **23** 4739–51

[55] Fuchs S *et al* 2009 Contribution of outgrowth endothelial cells from human peripheral blood on *in vivo* vascularization of bone tissue engineered constructs based on starch polycaprolactone scaffolds *Biomaterials* **30** 526–34

[56] Kim J Y and Cho D W 2009 Blended PCL/PLGA scaffold fabrication using multi-head deposition system *Microelectron. Eng.* **86** 1447–50

[57] Kundu J, Shim J-H, Jang J, Kim S-W and Cho D-W 2013 An additive manufacturing-based PCL–alginate–chondrocyte bioprinted scaffold for cartilage tissue engineering *Journal of Tissue Engineering and Regenerative Medicine* at press

[58] Lee J-S, Hong J M, Jung J W, Shim J-H, Oh J-H and Cho D-W 2014 3D printing of composite tissue with complex shape applied to ear regeneration *Biofabrication* **6** 024103

[59] Shim J-H *et al* 2013 Fabrication of blended polycaprolactone/poly (lactic-co-glycolic acid)/b-tricalcium phosphate thin membrane using solid freeform fabrication technology for guided bone regeneration *Tissue Eng. A* **19** 317–27

[60] Shim J-H, Lee J-S, Kim J Y and Cho D-W 2012 Bioprinting of a mechanically enhanced three-dimensional dual cell-laden construct for osteochondral tissue engineering using a multi-head tissue/organ building system *J. Micromech. Microeng.* **22** 085014

[61] Wiria F E, Leong K F, Chua C K and Liu Y 2007 Poly-ε-caprolactone/hydroxyapatite for tissue engineering scaffold fabrication via selective laser sintering *Acta Biomaterialia* **3** 1–12

[62] Peter S J, Miller M J, Yaszemski M J and Mikos A G 1997 Poly(propylene fumarate) *Handbook of Biodegradable Polymers* ed A J Domb, J Kost and D M Wiseman (Amsterdam: Harwood Academic) pp 87–97

[63] Lee J W, Kang K S, Lee S H, Kim J-Y, Lee B-K and Cho D-W 2011 Bone regeneration using a microstereolithography-produced customized poly(propylene fumarate)/diethyl fumarate photopolymer 3D scaffold incorporating BMP-2 loaded PLGA microspheres *Biomaterials* **32** 744–52

[64] Lin C C and Anseth K 2009 PEG hydrogels for the controlled release of biomolecules in regenerative medicine *Pharm. Res.* **26** 631–43

[65] Bryant S J, Davis-Arehart K A, Luo N, Shoemaker R K, Arthur J A and Anseth K S 2004 Synthesis and characterization of photopolymerised multifunctional hydrogels: water-soluble poly(vinyl alcohol) and chondroitin sulfate macromers for chondrocyte encapsulation *Macromolecules* **2004** 37

[66] Geckil H, Xu F, Zhang X, Moon S and Demirci U 2010 Engineering hydrogels as extracellular matrix mimics *Nanomedicine* **2010**(5) 469–84

[67] Klein G T, Lu Y and Wang M Y 2013 3D Printing and neurosurgery—ready for prime time? *World Neurosurg.* **80** 233–5

[68] Yoshii T *et al* 2010 A sustained release of lovastatin from biodegradable, elastomeric polyurethane scaffolds for enhanced bone regeneration *Tissue Eng.* **16** 2369–79

[69] Zhang C, Wen X, Vyavahare N R and Boland T 2008 Synthesis and characterization of biodegradable elastomeric polyurethane scaffolds fabricated by the inkjet technique *Biomaterials* **29** 3781–91

Dong-Woo Cho, Jung-Seob Lee, Falguni Pati, Jin Woo Jung, Jinah Jang and Jeong Hun Park

Chapter 8

Decellularized extracellular matrix (dECM)

The ECM provides important cues, with functional and structural molecules, for cellular behaviors [1]. Distinctive components in ECM provide complex mixtures, and the unique composition and topographic information regulates cell proliferation and phenotype during the tissue regeneration process. Fundamental interactions between the cells and matrix via membrane-bound integrin proteins determine cell lineages specific to a tissue site [2–4]. Various recent findings have indicated that tissue and organ-derived ECM are crucial for preserving cellular functions and phenotypes. It is known that cells can recognize their resident environment, migrate along protein footprints of previous cells and function similarly to previously resident cells. Specific binding sites in the matrix direct tissue development and maturation. Furthermore, the biochemical and mechanical signals residing in ECM are important in regulating cell fate by ECM-mediated stress and strain properties. In determining the success of tissue engineering grafts, it is important to choose the ideal material for obtaining physiological signals. Current research on decellularized ECM (dECM) highlights the need for tissue specificity to preserve a tissue homologous environment; this can provide benefits that non-tissue matched materials do not offer [1, 5, 6].

The use of dECM material was advanced notably by several pioneering studies on whole organ decellularization (e.g., heart, liver, and kidney) and its use as an underlying ECM scaffold for tissue engineering [7–9]. Later works demonstrated that dECM could be processed into an injectable hydrogel at physiological temperatures and injected *in vivo* [10]. Various applications have resulted in improved regeneration of tissues and induced recruitment of neighboring cells, and have demonstrated promising use as a novel cell or biomolecule delivery platform and biomaterial therapy [11]. There are various FDA-approved and clinically used dECM products, including skin [12], small intestine submucosa [13], pulmonary valves [14] and bone tissues derived from both allogeneic and xenogeneic tissues [15]. Alloderm is one of the well-known dECM products from human skin and it has

doi:10.1088/978-1-6817-4079-9ch8   8-1

been shown to be useful in multiple applications. Decellularized human valves (CryoValve®, CryoLife Inc.) are also currently implanted in right ventricular outflow tract reconstruction procedures. These dECM products can promote cell repopulation in the host and be maintained at a level similar to that of native tissue.

A recent advance in printable and tissue-specific dECM bioink has accelerated the field of 3D cell/organ printing, as used by Cho's group (figure 8.1(a)). The printable dECM bioinks derived from tissues, including adipose, heart and cartilage tissues, were developed and printed along with a PCL framework to support the 3D shape of tissue analogs [16]. In addition, Visser *et al* created crosslinkable hydrogels by covalent incorporation of dECM material from cartilage, meniscus, and tendon tissue in versatile and printable GelMA bioinks [17]. 3D printed adipose tissue constructs using decellularized adipose tissue bioink also resulted in positive tissue infiltration, constructive tissue remodeling, and adipose tissue formation (figure 8.1(b)) [5].

This chapter focuses on the processes for, and the characteristics of, dECM-derived hydrogels, which can be a precursor form of bioink; methods to modulate their material properties; and the significant progress in 3D printing applications. Additional studies have used the 3D native scaffolding material of whole decellularized organs/tissues, which may be a promising technique to regenerate whole organs; here, however, we focus only on the printable dECM materials containing native ECM material.

8.1 Preparation of decellularized ECM bioinks

To preserve the optimal function of dECM material, procedures for decellularization should be regulated. A decellularization protocol naturally disrupts certain portions of native ECM to provide a way to remove cellular debris. In particular, enzyme-based treatments digest much of the collagen-based ECM, and chemicals damage basement membrane proteins. Chemical-based treatment is known to provide favorable effects by removing antigens on the cell surface, as indicated by the ability to generate a pathogen-free material. dECM can be generated from various species and tissues (e.g., cadaveric, porcine and bovine tissues), and those tissues are minced into smaller pieces. Cell membranes are lysed using physical or chemical treatments, followed by separation of cellular components from the ECM using enzymatic treatments. Various decellularization techniques have also been described elsewhere, and a suitable protocol has to be selected depending on the tissue type.

The decellularization and sterilization methods can markedly influence the tissue remodeling and the functional outcome as well as the immune response. Keane *et al* evaluated the consequences of ineffective dECM material on the host immune response, and they suggested the cause of an inflammatory response might be immune rejection of xenogeneic DNA remaining in the dECM material [18]. Peracetic acid (PAA) treatment is an appropriate option for sterilization of dECM material and it removes bactericidal, fungicidal and sporicidal components from the dECM while preserving ECM proteins [19]. After complete decellularization, the

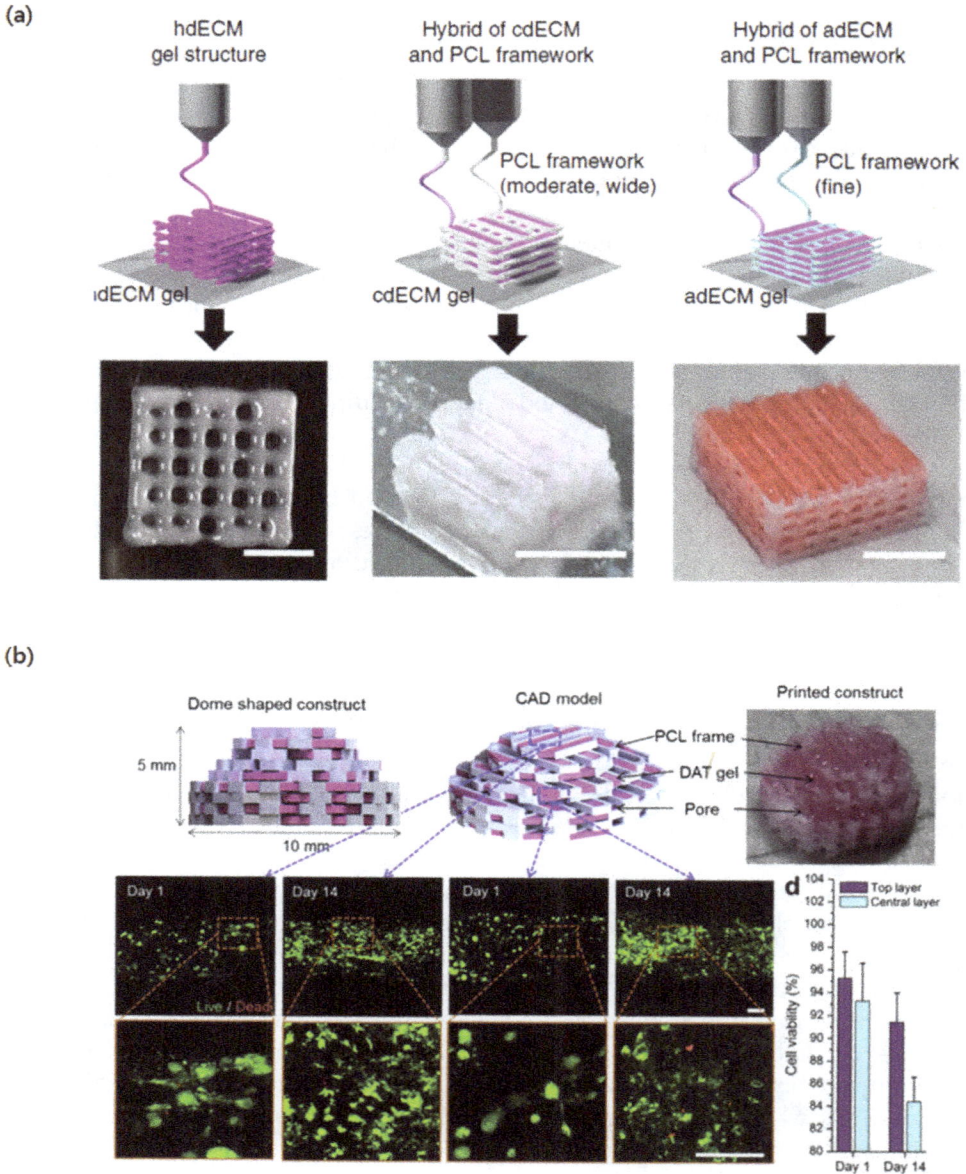

Figure 8.1. Printing process for particular tissue constructs with dECM bioink. (a) Heart tissue construct printed with only heart dECM (hdECM). Cartilage and adipose tissues were printed with cartilage dECM (cdECM) and adipose dECM (adECM), respectively, and in combination with a PCL framework (scale bar, 5 mm). (b) Dimensions of the tissue printed dome-shaped construct used for this study, and printed tissue constructs showing the PCL framework, DAT gel and pores, and the cell viability test results. Reproduced with permission from [5, 16].

tissues are extensively rinsed to remove detergent and then lyophilized, and the dECM is milled into a fine powder.

dECM powder is re-suspended and digested with pepsin, which can cleave the telopeptide regions of the collagen and allows the solublization of the dECM powder in acid solution. After digestion for 48 h, the pepsin is inactivated by adjusting the pH to 7.4 with NaOH solution. 10× PBS solution is added in the digested solution to match the ionic concentration and it is diluted to a desired concentration using 1× PBS solution. The prepared dECM bioink is then mixed with cells at a desired concentration or used itself for further printing processes. dECM bioink is loaded into the sterilized syringe and installed at the printing head; it should be maintained below 15 °C to avoid gelation during cell printing.

8.2 Properties of decellularized ECM bioinks

Histological and biochemical analyses are used widely to confirm the efficiency of the selected decellularization method. Hematoxylin and eosin (H&E) staining can be used as the first line of inspection to determine whether nuclear structures are observed after treatment. DNA quantification with Hoechst 33258, propidium iodide or PicoGreen is also used to obtain quantitative data regarding the presence of DNA within the dECM material. Although the definition of decellularization has not yet been clearly determined quantitatively, <50 ng double stranded DNA (dsDNA) per mg ECM dry weight and the lack of visible nuclear material in a histological analysis have been recommended for decellularized tissues [19]. In addition, the components of the ECM can be verified to determine what has been removed after decellularization. ECMs in the tissue, such as collagen, GAGs, laminin or fibronectin can be identified by spectroscopy-based quantification, antibody-based component-specific staining or enzyme-linked immunosorbent assay (ELISA) methods.

The rheological properties of dECM bioink are also relevant to the printability of the bioink. Rheological investigations are usually conducted to assess the steady or dynamic viscosity, frequency sweep and temperature sweep analysis, and these are often interlinked for the gelation kinetics, cell viability and shape fidelity of the printed construct. Pati *et al* tested the rheological properties of the dECM bioink. Their results showed that all dECM bioinks had shear thinning behavior and exhibited relative viscosities from 2.8 to 23.6, depending on the type of tissue. They measured viscosities using a rheometer below 15 °C to mimic the conditions for printing and used 3% dECM bioink. In addition, a temperature sweep analysis from 4 °C to 37 °C revealed that dECM bioink behaved like a thermally crosslinked hydrogel and exhibited greater storage than loss modulus with a 37 °C incubation for 30 min [16]. Most printing processes require bioinks with sufficient mechanical properties to withstand 3D structures, and printing fidelity generally increases with increasing viscosity and polymer concentration. However, these criteria imply an increase of the applied shear stress, which is harmful for cells in the bioink (figure 8.2). Researchers have now moved towards optimizing the parameters and processing conditions of bioink in a systematic way, with the intention of expanding the bioprinting window.

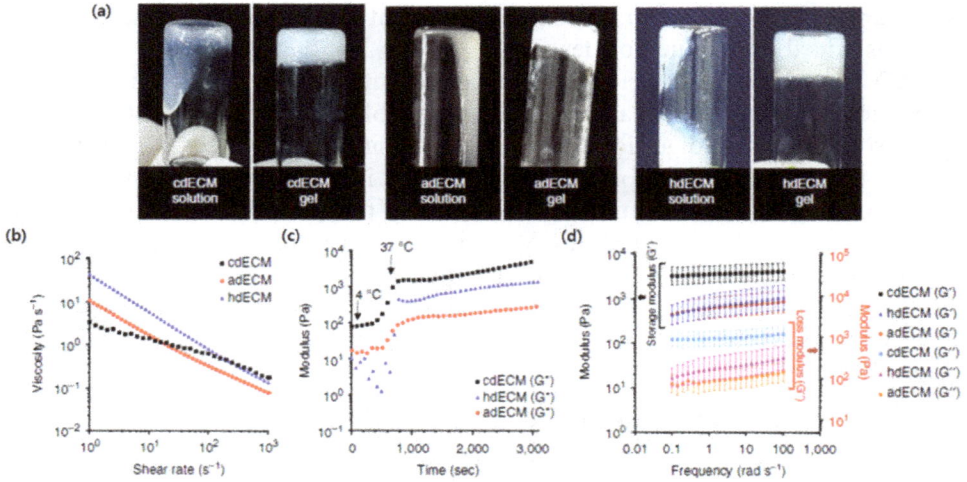

Figure 8.2. Rheological behavior of dECM pre-gels. Sol to gel transition (a) of the dECM pre-gels prepared from cartilage dECM (cdECM), heart dECM (hdECM) and adipose dECM (adECM). Rheological properties of the dECM pre-gels: (b) viscosity at 15 °C, (c) gelation kinetics from 4 °C to 37 °C (initial temperature: 4 °C, increment of 5 °C min^{-1} with holding at 15 °C for 5 min to reach 37 °C, maintained at 37 °C for 40 min) and (d) dynamic modulus at varying frequency at 37 °C. All experiments were performed in triplicate. Error bars represent SD. Reproduced with permission from [5].

8.3 Significant progress for 3D printing application

Various hydrogels have been used as bioinks for 3D cell printing and these are advantageous for cell culture because they are highly permeable to cell culture media, nutrients and waste generated during the metabolic activities of cells. They also have the ability to be fabricated in desired shapes, with tailorable material properties and various dimensions. dECM can be also processed to hydrogel-type materials, and various applications such as injectable hydrogel with cells, biomolecules and drugs have demonstrated the superiority of dECM materials for tissue regeneration. As we discussed earlier, dECM materials derived from heart, cartilage and fat tissue have been successfully processed into printable bioink and have shown a dramatic enhancement of the differentiation and maturation of the printed cells encapsulated within the developed bioink compared with collagen bioink.

As a follow-up study, Pati *et al* generated printed adipose tissue constructs using decellularized adipose tissue matrix (DAT) bioink encapsulating human adipose tissue-derived mesenchymal stem cells (hASCs). The constructs had dome shapes with porosity engineered by alternating printing with DAT bioink. By introducing this design parameter, high cell viability was facilitated over two weeks, and higher expression of representative adipogenic genes was induced with no additional adipogenic growth molecules. These results showed that it did not induce any chronic inflammation or cytotoxicity post-implantation, but supported positive tissue remodeling and adipose tissue formation [5].

Jang *et al* demonstrated an easy, versatile, and biocompatible two-step process that enabled tailoring of the physical and rheological properties of dECM bioink through mixing it with VB2 and exposing it to UVA light during the cell printing process. In particular, 0.02% VB2 mixed with 2% hdECM bioink exhibited a viscosity similar to that of the control hdECM bioink. However, after two-step crosslinking, the VB2-mixed hdECM gel was almost 33 times stiffer than the control hdECM bioink; this stiffness was similar to that of native cardiac tissue. The modified hdECM bioink was used to draw lines with various thicknesses by adjusting the printing parameters, which can facilitate a desired cellular arrangement by allowing control of the sizes of cellular clusters and their localization by 3D cell printing technology. It also provided increased mRNA expression of the genes (GATA4, Nkx 2.5, MEF2C, cTnI) related to cardiac differentiation due to the building of a cardiac tissue homologous microenvironment using the developed bioink.

References

[1] Crapo P M, Gilbert T W and Badylak S F 2011 An overview of tissue and whole organ decellularization processes *Biomaterials* **32** 3233–43

[2] Nelson C M and Bissell M J 2006 Of extracellular matrix, scaffolds, and signaling: tissue architecture regulates development, homeostasis, and cancer *Annu. Rev. Cell Dev. Biol.* **22** 287

[3] Guilak F, Cohen D M, Estes B T, Gimble J M, Liedtke W and Chen C S 2009 Control of stem cell fate by physical interactions with the extracellular matrix *Cell Stem Cell* **5** 17–26

[4] Ng S L, Narayanan K, Gao S and Wan A C 2011 Lineage restricted progenitors for the repopulation of decellularized heart *Biomaterials* **32** 7571–80

[5] Pati F, Ha D-H, Jang J, Han H H, Rhie J-W and Cho D-W 2015 Biomimetic 3D tissue printing for soft tissue regeneration *Biomaterials* **62** 164–75

[6] French K M *et al* 2012 A naturally derived cardiac extracellular matrix enhances cardiac progenitor cell behavior *in vitro Acta Biomaterialia* **8** 4357–64

[7] Ott H C *et al* 2010 Regeneration and orthotopic transplantation of a bioartificial lung *Nat. Med.* **16** 927–33

[8] Ott H C *et al* 2008 Perfusion-decellularized matrix: using nature's platform to engineer a bioartificial heart *Nat. Med.* **14** 213–21

[9] Song J J, Guyette J P, Gilpin S E, Gonzalez G, Vacanti J P and Ott H C 2013 Regeneration and experimental orthotopic transplantation of a bioengineered kidney *Nat. Med.* **19** 646–51

[10] Seif-Naraghi S B *et al* 2013 Safety and efficacy of an injectable extracellular matrix hydrogel for treating myocardial infarction *Sci. Trans. Med.* **5** 173

[11] Nguyen M M, Gianneschi N C and Christman K L 2015 Developing injectable nanomaterials to repair the heart *Curr. Opin. Biotech.* **34** 225–31

[12] Chen R-N, Ho H-O, Tsai Y-T and Sheu M-T 2004 Process development of an acellular dermal matrix (ADM) for biomedical applications *Biomaterials* **25** 2679–86

[13] Badylak S F *et al* 1995 The use of xenogeneic small intestinal submucosa as a biomaterial for Achilles tendon repair in a dog model *J. Biomed. Mater. Res.* **29** 977–85

[14] Elkins R C, Dawson P E, Goldstein S, Walsh S P and Black K S 2001 Decellularized human valve allografts *Ann. Thorac. Surg.* **71** S428–S32

[15] Benders KE, van Weeren P R, Badylak S F, Saris D B, Dhert W J and Malda J 2013 Extracellular matrix scaffolds for cartilage and bone regeneration *Trends Biotechnol.* **31** 169–76

[16] Pati F *et al* 2014 Printing three-dimensional tissue analogues with decellularized extracellular matrix bioink *Nat. Commun.* **2014** 5

[17] Visser J *et al* 2015 Crosslinkable hydrogels derived from cartilage, meniscus, and tendon tissue *Tissue Eng.* A **21** 1195–206

[18] Keane T J, Londono R, Turner N J and Badylak S F 2012 Consequences of ineffective decellularization of biologic scaffolds on the host response *Biomaterials* **33** 1771–81

[19] Gilbert T W, Sellaro T L and Badylak S F 2006 Decellularization of tissues and organs *Biomaterials* **27** 3675–83

Organ Printing

Dong-Woo Cho, Jung-Seob Lee, Falguni Pati, Jin Woo Jung, Jinah Jang and Jeong Hun Park

Chapter 9

Tissue engineering: bone

Bone is a complex composite tissue, consisting of different cell types, hydroxyapatite, collagen matrix and water. The major functions of bone are to provide a structural framework, mechanical strength, blood pH regulation and the maintenance of calcium and phosphate levels for metabolic processes [1, 2]. Bone tissue engineering based on 3D printing has been developed by studying the processability of Food and Drug Administration-approved biomaterials, and synthesizing new biomaterials to develop bone substitutes using an organ printing system [3]. Advances in 3D printing-based approaches have allowed bone substitutes to have not only sophisticatedly controlled inner architectures, but also the complex geometry of bone defects.

Bone substitutes have been developed using biomaterials that can provide mechanical and structural support during the regeneration of bone defects, while controlled delivery systems for growth factors, drugs and genes have been also incorporated into their fabrication, which can facilitate bone regeneration. Generally, an 'ideal' substitute for bone tissue engineering should satisfy the following diverse requirements [1, 4–6].

Biocompatibility is one of the most important requirements for bone substitutes. An ideal bone substitute should support cellular activities, such as cell proliferation, differentiation and extracellular matrix formation, without any adverse effect on the host tissue. Furthermore, an ideal bone substitute should have the ability to be incorporated within host bone, and to induce blood vessel formation inside and around the bone substitute within a few weeks of implantation to support nutrient, oxygen and waste transport [4].

Sufficient mechanical properties are important for bone substitute; they should be matched with those of the host bone and be able to withstand *in vivo* loading forces. The mechanical properties of bone vary considerably, from cancellous to cortical bone, and this large variation makes it difficult to tailor an ideal bone substitute. The mechanical properties of the bone substitute should also be designed according to the type of bone defect and fracture site.

A proper inner architecture is essential for the ideal substitute for bone tissue engineering. The bone substitute should have an interconnected porous architecture for sufficient diffusion of nutrients and oxygen for successful cellular activities, and pores sized 200–350 μm for effective ingrowth of bone tissue [7]. It has also been reported that a bone substitute with multi-scale pores can perform better than bone substitutes with only single-scale pores [8]. High porosity and interconnectivity between pores are important for bone tissue engineering applications; however, these are also directly related to the mechanical properties of bone substitute, and too high a porosity can reduce the compressive strength of the bone substitute.

Biodegradability is another key factor for the bone substitute. An ideal bone substitute should be able to degrade at a controlled rate in concert with the growth of new bone tissue over time to maintain structural integrity. Furthermore, the degradation products should not be toxic and should be easily excreted via metabolic pathways.

To date, diverse 3D printing techniques have been developed to create 3D interconnected porous bone substitutes or cell-laden constructs for bone tissue engineering. Lee *et al* used a photo-curable biomaterial, poly(propylene fumarate) (PPF)/diethyl fumarate (DEF), in a MSTL system [9, 10], and developed a 3D bone substitute that could facilitate bone regeneration by incorporating bone morphogenetic protein-2 (BMP-2)-loaded poly(DL-lactic-co-glycolic acid) (PLGA) microspheres into a 3D bone substitute [11]. The 3D printed bone substitute showed sustained release of BMP-2 *in vitro* for one month and *in vivo* results showed significantly enhanced bone formation in critical-size defects in rat cranial bone.

Seol *et al* used bioactive ceramic materials to develop a 3D bone substitute for bone tissue engineering [12]. They prepared a ceramic slurry by mixing hydrox-yapatite (HA) and tricalcium phosphate (TCP) powder with a photopolymer, and used a pMSTL system to print the 3D structure, consisting of ceramic powder and solidified photopolymer. A sintering process was then followed, not only to sinter the solidified photopolymer in the structure, but also to allow the ceramic particles to adhere to one another. The researchers also confirmed the sustained release of calcium ions from the ceramic scaffolds, and the effects of calcium ions on osteogenic differentiation and bone tissue regeneration were demonstrated by *in vitro* and *in vivo* experiments [13].

Shim *et al* used a dispensing-type 3D printing system and developed a hybrid substitute consisting of synthetic biomaterials and natural hydrogel for bone tissue engineering [14]. They fabricated a framework using a blend of polycaprolactone (PCL) and poly (lactic-coglycolic acid) (PLGA)), and natural hydrogel solutions, such as atelocollagen, hyaluronic acid (HA), and gelatin were infused into the canals of a framework. In an *in vitro* test, the hybrid substitute showed a much higher cell affinity than the PCL/PLGA substitute alone. They also developed a PCL/PLGA substitute blended with TCP, and its ability to form new bone was demonstrated in *in vivo* and *in vitro* experiments [15, 17].

Shim *et al* also developed a growth factor delivery system in a 3D printing based PCL/PLGA substitute for enhanced bone tissue regeneration [18]. They dispensed recombinant human BMP-2 (rhBMP-2) with a collagen hydrogel into a 3D printed hollow cylindrical PCL/PLGA substitute for long-term delivery (up to a month), and also dispensed gelatin hydrogel in the same way for short-term delivery (within a week). In an *in vivo* test, the long-term release of rhBMP-2 by the PCL/PLGA/collagen/rhBMP-2 substitute showed significantly accelerated new bone formation without significant inflammatory response in 20 mm rabbit radius defects.

Park *et al* developed a pre-vascularized bone tissue as a rational strategy to overcome the size limitation of tissue implants and to enhance bone tissue regeneration using 3D cell/organ printing technology [19]. They confirmed hypoxic area formation in the large volume structure by evaluating the cell survival rate. They then printed mesenchymal dental pulp-derived stem cells (DPSCs) with vascular endothelial growth factor (VEGF) in the central zone, in which a hypoxic area formed, and printed DPSCs with BMP-2 in the peripheral zone of the 3D structure. In their *in vitro* and *in vivo* tests, 3D cell-printed structures with DPSCs and two kinds of growth factors showed sufficient blood vessel formation, and facilitated bone formation even in the core of a large structure.

References

[1] Venkatesan J, Bhatnagar I, Manivasagan P, Kang K H and Kim S K 2015 Alginate composites for bone tissue engineering: a review *Int. J. Biol. Macromol.* **72** 269–81

[2] Sowjanya J A, Singh J, Mohita T, Sarvanan S, Moorthi A, Srinivasan N and Selvamurugan N 2013 Biocomposite scaffolds containing chitosan/alginate/nano-silica for bone tissue engineering *Colloids Surf.* B **109** 294–300

[3] Seol Y J, Kang T Y and Cho D W 2012 Solid freeform fabrication technology applied to tissue engineering with various biomaterials *Soft Matter.* **8** 1730–35

[4] Bose S, Roy M and Bandyopadhyay A 2012 Recent advances in bone tissue engineering scaffolds *Trends Biotechnol.* **30** 546–54

[5] Liu Y, Lim J and Teoh S H 2013 Review: development of clinically relevant scaffolds for vascularized bone tissue engineering *Biotechnol. Adv.* **31** 688–705 PMID: 3534809

[6] Liu X and Ma P X 2004 Polymeric scaffolds for bone tissue engineering *Ann. Biomed. Eng.* **32** 477–86

[7] Murphy C M, Haugh M G and O'Brien F J 2010 The effect of mean pore size on cell attachment, proliferation and migration in collagen–glycosaminoglycan scaffolds for bone tissue engineering *Biomaterials* **31** 461–6

[8] Woodard J R *et al* 2007 The mechanical properties and osteoconductivity of hydroxyapatite bone scaffolds with multi-scale porosity *Biomaterials* **28** 45–54

[9] Lee J W, Lan P X, Kim B, Lim G and Cho D W 2008 Fabrication and characteristic analysis of a poly(propylene fumarate) scaffold using micro-stereolithography technology *J. Biomed. Mater. Res.* B **87** 1–9

[10] Lan P X, Lee J W, Seol Y J and Cho D W 2009 Development of 3D PPF/DEF scaffolds using micro-stereolithography and surface modification *J. Mater. Sci.: Mater. Med.* **20** 271–9

[11] Lee J W, Kang K S, Lee S H, Kim J Y, Lee B K and Cho D W 2011 Bone regeneration using a microstereolithography-produced customized poly(propylene fumarate)/diethyl fumarate photopolymer 3D scaffold incorporating BMP-2 loaded PLGA microspheres *Biomaterials* **32** 744–52

[12] Seol Y J, Park D Y, Park J Y, Kim S W, Park S J and Cho D W 2012 A new method of fabricating robust freeform 3d ceramic scaffolds for bone tissue regeneration *Biotechnol. Bioeng.* **110** 1444–55

[13] Seol Y J, Park D Y, Jung J W, Jang J, Girdhari R, Kim S W and Cho D W 2014 Improvement of bone regeneration capability of ceramic scaffolds by accelerated release of their calcium ions *Tissue Eng.* A **20** 2840–9

[14] Shim J H, Kim J Y, Park M, Park J and Cho D W 2011 Development of a hybrid scaffold with synthetic biomaterials and hydrogel using solid freeform fabrication technology *Biofabrication* **3** 034102

[15] Shim J H, Moon T S, Yun M J, Jeon Y C, Jeong C M, Cho D W and Huh J B 2012 Stimulation of healing within a rabbit calvarial defect by a PCL/PLGA scaffold blended with TCP using solid freeform fabrication technology *J. Mater. Sci: Mater. Med.* **23** 2993–3002

[16] Shim J H, Huh J B, Park J Y, Jeon Y C, Kang S S, Kim J Y, Rhie J W and Cho D W 2014 Fabrication of blended polycaprolactone/poly(lactic-co-glycolic acid)/β-tricalcium phosphate thin membrane using solid freeform fabrication technology for guided bone regeneration *Tissue Eng.* A **19** 317–27

[17] Shim J H, Yoon M C, Jeong C M, Jang J, Jeong S I, Cho D W and Huh J B 2014 Efficacy of rhBMP-2 loaded PCL/PLGA/β-TCP guided bone regeneration membrane fabricated by 3D printing technology for reconstruction of calvaria defects in rabbit *Biomed. Mater.* **9** 065006

[18] Shim J H, Kim S E, Park J Y, Kundu J, Kim S W, Kang S S and Cho D W 2014 Three-dimensional printing of rhbmp-2-loaded scaffolds with long-term delivery for enhanced bone regeneration in a rabbit diaphyseal defect *Tissue Eng.* A **20** 1980–92

[19] Park J Y, Shim J H, Choi S A, Jang J, Kim M, Lee S H and Cho D W 2015 3D printing technology to control BMP-2 and VEGF delivery spatially and temporally to promote large-volume bone regeneration *J. Mater. Chem.* B **3** 5415–25

Organ Printing

Dong-Woo Cho, Jung-Seob Lee, Falguni Pati, Jin Woo Jung, Jinah Jang and Jeong Hun Park

Chapter 10

Tissue engineering: cartilage (nose, ear and trachea)

Cartilage tissue is a thin and avascular tissue, elegantly organized to allow for elastic and smooth motion. The bodies of humans and other animals possess various cartilaginous tissues in many areas, including the nose, ears, joints, airway, intervertebral disks and ribs. Chondrocytes are sparsely distributed within the ECM of cartilage tissue, composed mainly of collagens and proteoglycans. This matrix provides sufficient mechanical properties to withstand repetitive mechanical stresses and to minimize friction between the joints for effortless motion. However, aging, disease and injury to the cartilage tissue lead to further degeneration over its lifetime [1]. Additionally, due to its avascularity, cartilage has an innately limited capacity for self-repair.

Tissue engineering approaches are considered to be a promising strategy to repair cartilage defects. Using a scaffold and/or a hydrogel containing an appropriate cell source and biological molecules, many researches have shown somewhat encouraging results with regenerated neo-cartilage tissue, with cartilage-like matrix formation. For a decade, 3D printing has been used to mimic the 3D structure of a specific native cartilage tissue, such as the meniscus [2], or for cosmetic purposes, such as ear [3] and nasal implants [4]. The aim of this chapter is to provide an overview of the necessary background and current trends in 3D printing for cartilage tissue engineering.

10.1 3D printing for cartilage tissue regeneration

Due to the poor self-repairing ability of cartilage tissue, existing healthy cartilage should be preserved in repair processes. Thus, customized scaffold fabrication using 3D printing has been an attractive option for close replacement of a cartilage lesion without additional removal of any healthy region [5]. The implanted structure with an ideal shape would be expected to form native cartilage-like tissue, well integrated with the existing native cartilage. Despite great interest in cartilage for tissue engineering, there are still several challenges that should be addressed, such as cell sources and biomaterials selection, and construction of the optimal scaffold. Each of these issues is discussed in detail.

10.1.1 Cell source selection

The selection of the optimal cell type for cartilage tissue regeneration remains a challenge. Production of cartilage tissue-specific ECM is the most important factor, and cell differentiation towards the desired phenotype has to be considered. With these criteria, mesenchymal stem cells (MSCs) and autologous chondrocytes have been the best cell candidates for cartilage therapies [6]. Non-expanded autologous chondrocytes have an ideal chondrocytic phenotype, but there are issues with the limited quantity of donor tissue and early degeneration [7]. Using MSCs for cartilage regeneration is a promising strategy to avoid the problems of chondrocytes, although MSCs need to differentiate into a chondrogenic lineage [6]. Recent studies have shown that MSC and chondrocyte cocultures could be an effective approach to obtain well-organized cartilage tissue by stimulating MSC chondrogenesis [8, 9].

10.1.2 Biomaterial selection

Various biomaterials have been used for scaffolds in cartilage tissue engineering. The ideal scaffold should have (i) a biodegradation rate matched with the new tissue formation rate, (ii) biological components for the chondrogenic differentiation of the contained cells, (iii) an appropriate supply of nutrients and oxygen for the cells and (iv) mechanical properties that protect cells from external forces. Cartilage ECM is composed of 50–75% collagen and 15–30% glycosaminoglycans (GAGs) [10]. In this regard, collagen-based and collagen-like hydrogels (e.g., type 1 or 2 collagen, gelatin, chitosan and decellularized cartilage ECM [11, 12]) have been used to mimic the biological and chemical milieu of native cartilage tissue. The hydrogels have 3D hydrophilic networks that can contain 10–20% water, allowing the diffusion of solutes and nutrients into cells to maintain cell viability. In addition, the hydrogels provide a 3D cell–cell and cell–matrix interaction microenvironment for efficient chondrogenic differentiation of MSCs or chondrocytes [13].

However, the crosslinked structure has poor mechanical strength for maintaining physiological or external loads over the long term [14]. The implanted structure has to withstand physiological stresses of over 2 MPa for cartilage tissue regeneration [12]. Several biodegradable synthetic polymers, such as poly(L-lactide) (PLLA), poly(glycolide) (PGA), poly(DL-lactide-co-glycolide) (PLGA) and poly(ε-caprolactone) (PCL), have been used to provide better physiological and biomechanical properties. These materials can be used for forming desired shapes, and the mechanical properties and biodegradation rate can be manipulated readily according to the specific target tissues. However, additional biochemical treatments or immobilization of hydrogels on these synthetic-derived polymers may be required for better forming of cartilage tissue, because they lack chondrogenic-specific bio-components [15].

10.1.3 Scaffold fabrication using 3D printing for cartilage regeneration

In cartilage tissue engineering, various biocompatible materials have been used for the creation of tissue-engineered scaffolds that are physically or biochemically supplemented with additional cartilage-specific components, such as growth factors [16], hydrogels [17] and bio-stimuli [18]. These studies have contributed to cartilage tissue engineering, becoming a promising therapeutic method for the treatment of cartilage lesions. However, despite decades of research and many reports, most of the approaches for cartilage repair are still far from providing acceptable clinical results and generating a tissue comparable with native cartilage in elastic modulus, strength and friction coefficient.

For clinical success, a better understanding of clinical issues with respect to cells, growth factors, materials and scaffold processing is required. With no chemical modification, physical combinations of clinically approved components might be available for clinical use. Considerable advances in 3D printing techniques for tissue engineering have been made recently, enabling the fabrication of a hybrid scaffold composed of synthetic polymer as a supporting frame and a cell-encapsulated hydrogel [19]. As mentioned above, the hydrogel part has good biological properties for cartilage tissue formation, and the synthetic polymer part provides additional benefits, supporting the original shape and controlling the biodegradation rate and mechanical strength. Kundu *et al* showed that a PCL-alginate-chondrocyte scaffold, printed using dispensing-based 3D printing with a multi-nozzle, promoted the secretion of GAG and cartilage-specific ECM *in vivo* [20]. Kim *et al* printed an ear-shaped scaffold composed of a PCL frame and two alginate hydrogels using four heads of a multi-nozzle printing system [3]. The materials used in these studies can be replaced conveniently with clinically approved biomaterials that will be acceptable for clinical studies. These materials, in a liquid state, are inserted in nozzles and then selectively positioned using a computer-aided control system. In future studies, this printing approach will be applied to the fabrication of new hybrid scaffolds composed of multiple materials, inspired by previous results for effective cartilage tissue formation, and their physically combinatorial effects and clinical effectiveness will be investigated using an animal model.

10.2 Conclusion

Native cartilage tissue has limitations with regard to spontaneous repair because it lacks proliferation potential in the chondrocytes, and there is little immigration of regenerative cells. Similarly, the regenerated cartilage tissue in the body will face the same harsh environment as native cartilage. Thus, the long-term clinical success of cartilage repair remains a major challenge in the tissue-engineering field, despite many successful results using animal models. Future research in cartilage tissue engineering should thus include the establishment of new strategies through the combination of current effective methods with a better understanding of clinical issues.

References

[1] O'Driscoll S W 1998 The healing and regeneration of articular cartilage *J. Bone Joint Surg. Am.* **80** 1795–812 PMID: 9875939

[2] Lee C H, Rodeo S A, Fortier L A, Lu C Y, Erisken C and Mao J J 2014 Protein-releasing polymeric scaffolds induce fibrochondrocytic differentiation of endogenous cells for knee meniscus regeneration in sheep *Sci. Transl. Med.* **2014** 6

[3] Lee J S, Hong J M, Jung J W, Shim J H, Oh J H and Cho D W 2014 3D printing of composite tissue with complex shape applied to ear regeneration *Biofabrication* **6** 024103

[4] Jung J W, Park J H, Hong J M, Kang H W and Cho D W 2014 Octahedron pore architecture to enhance flexibility of nasal implant-shaped scaffold for rhinoplasty *Int. J. Precis. Eng. Man.* **15** 2611–6

[5] Cui X F, Breitenkamp K, Finn M G, Lotz M and D'Lima D D 2012 Direct human cartilage repair using three-dimensional bioprinting technology *Tissue Eng.* A **18** 1304–12

[6] Vinatier C *et al* 2009 Cartilage tissue engineering: towards a biomaterial-assisted mesenchymal stem cell therapy *Current Stem Cell Research and Therapy* **4** 318–29

[7] Hui J H, Chen F, Thambyah A and Lee E H 2004 Treatment of chondral lesions in advanced osteochondritis dissecans: a comparative study of the efficacy of chondrocytes, mesenchymal stem cells, periosteal graft and mosaicplasty (osteochondral autograft) in animal models *J. Pediatr. Orthop.* **24** 427–33

[8] Bian L, Zhai D Y, Mauck R L and Burdick J A 2011 Coculture of human mesenchymal stem cells and articular chondrocytes reduces hypertrophy and enhances functional properties of engineered cartilage *Tissue Eng.* A **17** 1137–45

[9] Acharya C *et al* 2012 Enhanced chondrocyte proliferation and mesenchymal stromal cells chondrogenesis in coculture pellets mediate improved cartilage formation *J. Cell Physiol.* **227** 88–97

[10] Athanasiou K A, Darling E M and Hu J C 2009 Articular Cartilage tissue engineering *Syn. Lect. Tissue Eng.* **1** 1–182

[11] Pati F *et al* 2014 Printing three-dimensional tissue analogues with decellularized extracellular matrix bioink *Nat. Commun.* **2014** 5

[12] Balakrishnan B and Banerjee R 2011 Biopolymer-based hydrogels for cartilage tissue engineering *Chem. Rev.* **111** 4453–74

[13] Mackay A M, Beck S C, Murphy J M, Barry F P, Chichester C O and Pittenger M F 1998 Chondrogenic differentiation of cultured human mesenchymal stem cells from marrow *Tissue Eng.* **4** 415–28

[14] Kim B S and Mooney D J 1998 Development of biocompatible synthetic extracellular matrices for tissue engineering *Trends Biotechnol.* **16** 224–30

[15] Chen G P, Sato T, Ushida T, Ochiai N and Tateishi T 2004 Tissue engineering of cartilage using a hybrid scaffold of synthetic polymer and collagen *Tissue Eng.* **10** 323–30

[16] Park J H *et al* 2015 A novel tissue-engineered trachea with a mechanical behavior similar to native trachea *Biomaterials* **62** 106–15

[17] Chih-Hao C, Victor Bong-Hang S, Jyh-Ping C and Ming-Yih L 2014 Selective laser sintered poly-ε-caprolactone scaffold hybridized with collagen hydrogel for cartilage tissue engineering *Biofabrication* **6** 015004

[18] Gemmiti C V and Guldberg R E 2006 Fluid flow increases type II collagen deposition and tensile mechanical properties in bioreactor-grown tissue-engineered cartilage *Tissue Eng.* **12** 469–79

[19] Shim J H, Lee J S, Kim J Y and Cho D W 2012 Bioprinting of a mechanically enhanced three-dimensional dual cell-laden construct for osteochondral tissue engineering using a multi-head tissue/organ building system *J. Micromech. Microeng.* **2012** 22

[20] Kundu J, Shim J H, Jang J, Kim S W and Cho D W 2013 An additive manufacturing-based PCL-alginate-chondrocyte bioprinted scaffold for cartilage tissue engineering *J. Tissue Eng. Regen. Med.* (doi: 10.1002/term.1682)

[21] Kim K-J *et al* 2014 Role of pigment epithelium-derived factor in the involution of hemangioma: autocrine growth inhibition of hemangioma-derived endothelial cells *Biochem. Biophys. Res. Commun.* **454** 282–8

IOP Concise Physics

Organ Printing

Dong-Woo Cho, Jung-Seob Lee, Falguni Pati, Jin Woo Jung, Jinah Jang and Jeong Hun Park

Chapter 11

Tissue engineering: osteochondral tissue

Osteochondral tissue is composed of articular cartilage and underlying subchondral bone in the knee joint and is a complex heterogeneous tissue [1]. Each year, more than 12 million people visit the hospital because of knee pain and half of them suffer from damage to the articular cartilage [2]. Articular cartilage that covers the ends of bones in the knee joint is usually damaged in connection with defects or damage to the subchondral bone. Such a complex defect, composed of cartilage and bone, is called an osteochondral defect. Osteochondral defects are associated with mechanical instability of the joint with degenerative osteoarthiritis [3].

In current clinical research, the most common approach is to implant autologous osteochondral grafts obtained from a donor in a defect to restore the cartilage and bone. In this case, there are limitations in terms of donor supply and the donor site transformation, and the topology of the harvested grafts is not the same as that of the defect [4]. Various concepts and strategies to fabricate and restore the tissue engineered osteochondral defects have been reported. For example, heterogeneous scaffolds, made of two distinct but integrated layers for cartilage and bone regions, were fabricated for osteochondral defects [5]. The top layer was made of biomaterials, such as PLA, hyaluronic acid and collagen type 2 for the regeneration of native articular cartilage, and the bottom layer was composed of hydroxyapatite (HA) and tricalcium phosphate for native bone regeneration [6–8].

In these examples, 3D printing technology has advantages, particularly in the fabrication and reconstruction of the tissue-engineered osteochondral tissue, because it is a complex and heterogeneous tissue. In this chapter, we introduce the representative characteristics of and studies regarding the fabrication of osteochondral tissue using 3D printing. We first consider the essential requirements to fabricate tissue-engineered osteochondral tissue using 3D printing.

Biocompatibility is one of the most important requirements for tissue regeneration. Osteochondral tissue contains articular cartilage and bone. Articular cartilage is

doi:10.1088/978-1-6817-4079-9ch11

hyaline cartilage, containing chondrocytes, abundant collagen type 2, proteoglycans (such as aggrecan, GAGs and hyaluronic acid). Bone contains osteoblasts, abundant collagen type 1, and hydroxyapatite composites, with calcium and phosphate [9, 10]. For the regeneration of osteochondral tissue, the environment in the scaffold should have no toxicity issues, and should have similar components and assistance for the proliferation and differentiation of chondrocytes and osteoblasts.

Sufficient mechanical properties are also important for osteochondral scaffolds, to endure the loading forces *in vivo*. The osteochondral scaffold should withstand external force from walking movements and keep its 3D shape while chondrocytes and osteoblasts form cartilage and bone. Thus, many researchers have focused on the compressive modulus and strength of osteochondral scaffolds [11, 12].

An appropriate inner architecture is also required in a structure fabricated by 3D printing. The structure should have interconnected pores to allow the exchange of nutrients and oxygen for printed cells, for complete tissue formation and regeneration of the normal organization. Native chondrocytes in cartilage live at a low oxygen level and the native tissue is relatively thin [13–15]. However, an increase in oxygen may lead to abnormal levels of reactive oxygen species, which may result in disturbances to chondrocyte metabolism [16]. Additionally, the bottom bone part of an osteochondral structure should have adequately interconnected pores to supply nutrients and oxygen and, in particular, multi-scale pores can improve osteogenesis [17].

Biodegradability is also an important property in tissue-engineered osteochondral structures. Printed cells in hydrogels or synthetic polymers proliferate and differentiate to form target tissue in the *in vivo* environment; the natural and synthetic biomaterials should degrade adequately. If the degradation time is too short compared with the formation time of the target tissue, the scaffold may collapse and form an abnormal shape [18]. As already mentioned, articular cartilage and subchondral bone in osteochondral tissue have different inherent characteristics. The choice of biomaterials for use with cells should take into consideration the periods taken by chondrocytes and osteoblasts to differentiate and form cartilage and bone, respectively.

Biomaterial and cell sources are also important in osteochondral tissue regeneration with structures fabricated using 3D printing. Bone and cartilage reconstruction or regeneration by autologous cell/tissue transplantation is one of the most promising technologies [19]. Various cells, such as osteoblasts, chondrocytes and mesenchymal stem cells obtained from the patient's hard and soft tissues, are generally used [18, 20]. Recently, adipose-derived stem cells (ASCs) have also attracted attention because they have the potential to differentiate into chondrocytes and osteoblasts [21–24]. Appropriate scaffold biomaterials to generate cartilage and bone tissue have been investigated, including hydroxyapatite (HA), polycaprolactone (PCL), poly(lactide-co-glycolide acid) (PLGA), and natural polymers, such as collagen type 1, hyaluronic acid (HA) and chitin [6–8, 18]. Recently, decellularized extracellular matrix (dECM) material obtained from the patient's own native tissue resulted in more effective bone and cartilage formation [25].

The development of fabricated structures and cell-printed structures with bio-materials using 3D printing have been reported for osteochondral tissue regeneration [6–8, 26, 28]. For such structures, many requirements were considered, such as biomimicry of osteochondral tissue, biocompatibility, mechanical properties, inner pore architecture and biomaterial and cell sources. 3D printing has the advantage that one or multiple cell types or cell-laden hydrogels can be printed in accurate positions [26–28], with completely interconnected pores in the scaffold and with geometric and structural enhancement of the mechanical compressive modulus and strength [29, 30].

Schek [8] *et al* fabricated a porous scaffold with a disc shape using conventional methods and 3D printing technology. The scaffold was composed of two parts, a poly(L-lactic acid) (PLA) sponge and a HA scaffold, for osteochondral tissue regeneration. The upper part, with a sponge-type structure (diameter 5 mm and height 3 mm), was made with PLA using a solvent-casting particulate-leaching method, and the bottom part was a HA scaffold (diameter 5 mm and height 3 mm), fabricated with a HA slurry and porous mold, manufactured by 3D printing. Polyglycolic acid (PGA) was used to glue the PLA sponge structure to the HA scaffold in fabricating the final PLA–HA structure. The porcine articular chondro-cyte and human gingival fibroblasts were seeded into the PLA structure and HA scaffold, respectively, and these were implanted in mice for 4 weeks. The study confirmed the presence of GAG-rich ECMs (in the cartilage), the blood vessels, the marrow stroma and adipose tissue (in the bone).

Cao *et al* [31] fabricated an osteochondral composites scaffold with PCL using FDM. The PCL scaffold has a honeycomb-like geometry (in a 0°/60°/120° lay-down pattern), and human bone marrow-derived mesenchymal progenitor cells and human rib chondrocytes were seeded in the scaffold, with the former in one half and the latter in the other half. The scaffolds seeded with the two kinds of cells were cultured for eight weeks and were evaluated to confirm chondrogenesis and osteo-genesis. They observed different extracellular matrices in each part; however, they did not perform further assays of matrix composition or cell phenotype.

Shao *et al* [6] fabricated medical-grade PCL (mPCL) scaffolds for the bone compartment with a FDM process, and fibrin glue was used as the matrix for the cartilage part. The mPCL structure was fabricated with PCL using FDM ($40 \times 40 \times 4$ mm^3 size) and the pore morphology was a porous structure like a honeycomb. They were stamped out with a skin biopsy punch (diameter 4 mm, height 4 mm) and bone marrow-derived mesenchymal cells (BMSC) were seeded in the disk scaffold. The cell–scaffold constructs were implanted in rabbit knee defects and the remaining cartilage defect was filled with fibrin containing BMSC. *In vivo* tests were conducted for three and six months and these evaluated osteogenesis and chondrogenesis. Assessing the *in vivo* results, it was concluded that mPCL scaffolds are a promising matrix for osteochondral bone regeneration, but fibrin glue is not a suitable scaffold for the reconstruction of articular cartilage at a weight-bearing site.

Seol *et al* [32] fabricated a hybrid scaffold made of a ceramic scaffold and an alginate/chondrocyte hydrogel. The ceramic scaffold (diameter 4 mm, height 6 mm) was porous and made of HA with laser-based 3D printing for bone formation. Before the upper alginate/chondrocyte hydrogel for cartilage generation was cross-linked, chondrocytes obtained with rabbit TGF-β growth factor treatment were mixed in 4% w/v alginate solution, and the alginate/chondrocytes were then crosslinked with CaCl$_2$ using a Teflon mold. In the crosslinking process, the HA scaffold was positioned directly and fixed to assemble the scaffold with the alginate/chondrocyte hydrogel at 1 mm in depth. The hybrid scaffolds were implanted in the knee joint of rabbits for 12 weeks and evaluated for osteogenesis and chondrogenesis. It was confirmed that the pores and HA components in the ceramic scaffold of the hybrid scaffold induced osteogenesis, but little chondrogenesis was seen in the alginate hydrogel due to the inadequate mechanical properties of the hydrogel.

Fedorovich *et al* [33] fabricated cell-laden hydrogel-based constructs with 3D fiber deposition (3DF). They encapsulated HMSC with osteoinductive biphasic calcium phosphate particles (BCPs) and human articular chondrocytes in an alginate hydrogel solution. A HMSC/BCPs-laden alginate hydrogel and a chondrocyte-laden alginate hydrogel were dispensed to fabricate the hydrogel construct with a rectangular shape ($10 \times 20 \times 0.8$ mm^3) using 3DF. They conducted *in vitro* and *in vivo* tests with nude mice (six weeks) using simple hydrogel constructs and evaluated osteogenesis and chondrogenesis. They concluded that the cell-laden hydrogel formed cartilage and bone and demonstrated the possibility of manufacturing viable heterogeneous tissue constructs with 3DF.

Shim *et al* [28] fabricated cell-printed structures for osteochondral tissue generation using a multi-head tissue/organ building system (MtoBS), a 3D cell printing technology. The cell-printed structure consisted of different kinds of cells. They first fabricated a PCL framework with pores and printed a prepared cell-laden hydrogel into secondary pores. Through a layer-by-layer process, they fabricated the final cell-printed structure, corresponding to cartilage and bone tissue. They proposed that it might be possible to regenerate heterogeneous tissue, such as osteochondral tissue, with cell-printed structures fabricated with MtoBS.

References

[1] Prakash D and Learmonth D 2002 Natural progression of osteo-chondral defect in the femoral condyle *The Knee* **9** 7–10

[2] Alonso J E, Lee J, Burgess A R and Browner B D 1996 The management of complex orthopedic injuries *Surg. Clin. N. Am.* **76** 879–903

[3] Tokuhara Y, Wakitani S, Imai Y, Kawaguchi A, Fukunaga K, Kim M, Kadoya Y and Takaoka K 2010 Repair of experimentally induced large osteochondral defects in rabbit knee with various concentrations of *Escherichia coli*-derived recombinant human bone morphogenetic protein-2 *Int. Orthop.* **34** 761–7

[4] Martin I, Miot S, Barbero A, Jakob M and Wendt D 2007 Osteochondral tissue engineering *J. Biomech.* **40** 750–65

[5] Sherwood J K, Riley S L, Palazzolo R, Brown S C, Monkhouse D C, Coates M, Griffith L G, Landeen L K and Ratcliffe A 2002 A three-dimensional osteochondral composite scaffold for articular cartilage repair *Biomaterials* **23** 4739–51

[6] Shao X X, Hutmacher D W, Ho S T, Goh J C and Lee E H 2006 Evaluation of a hybrid scaffold/cell construct in repair of high-load-bearing osteochondral defects in rabbits *Biomaterials* **27** 1071–80

[7] Solchaga L A, Gao J, Dennis J E, Awadallah A, Lundberg M, Caplan A I and Goldberg V M 2002 Treatment of osteochondral defects with autologous bone marrow in a hyaluronan-based delivery vehicle *Tissue Eng.* **8** 333–47

[8] Schek R M, Taboas J M, Segvich S J, Hollister S J and Krebsbach P H 2004 Engineered osteochondral grafts using biphasic composite solid free-form fabricated scaffolds *Tissue Eng.* **10** 1376–85

[9] Hrabchak C, Rouleau J, Moss I, Woodhouse K, Akens M, Bellingham C, Keeley F, Dennis M and Yee A 2010 Assessment of biocompatibility and initial evaluation of genipin cross-linked elastin-like polypeptides in the treatment of an osteochondral knee defect in rabbits *Acta Biomaterialia* **6** 2108–15

[10] Karageorgiou V and Kaplan D 2005 Porosity of 3D biomaterial scaffolds and osteogenesis *Biomaterials* **26** 5474–91

[11] Ikeda R *et al* 2009 The effect of porosity and mechanical property of a synthetic polymer scaffold on repair of osteochondral defects *Int. Orthop.* **33** 821–8

[12] Moroni L, De Wijn J R and Van Blitterswijk C A 2006 3D fiber-deposited scaffolds for tissue engineering: influence of pores geometry and architecture on dynamic mechanical properties *Biomaterials* **27** 974–85

[13] Clark C C, Tolin B S and Brighton C T 1991 The effect of oxygen tension on proteoglycan synthesis and aggregation in mammalian growth plate chondrocytes *J. Orthop. Res.* **9** 477–84

[14] Malda J *et al* 2004 Effect of oxygen tension on adult articular chondrocytes in microcarrier bioreactor culture *Tissue Eng.* **10** 987–94

[15] Schneider N *et al* 2007 Oxygen consumption of equine articular chondrocytes: influence of applied oxygen tension and glucose concentration during culture *Cell Biol. Int.* **31** 878–86

[16] Kemppainen J M and Hollister S J 2010 Differential effects of designed scaffold permeability on chondrogenesis by chondrocytes and bone marrow stromal cells *Biomaterials* **31** 279–87

[17] Murphy C M, Haugh M G and O'Brien F J 2010 The effect of mean pore size on cell attachment, proliferation and migration in collagen–glycosaminoglycan scaffolds for bone tissue engineering *Biomaterials* **31** 461–6

[18] Hutmacher D W 2000 Scaffolds in tissue engineering bone and cartilage *Biomaterials* **21** 2529–43

[19] Patrick C W Jr, Mikos A G and McIntire L V 1998 Prospectus of tissue engineering *Front. Tissue Eng.* **3** 14

[20] Langer R and Vacanti J P 1993 Tissue engineering *Science* **260** 920

[21] Huang J I, Zuk P A, Jones N F, Zhu M, Lorenz H P, Hedrick M H and Benhaim P 2004 Chondrogenic potential of multipotential cells from human adipose tissue *Plast. Reconstr. Surg.* **113** 585–94

[22] Strem B M, Hicok K C, Zhu M, Wulur I, Alfonso Z, Schreiber R E, Fraser J K and Hedrick M H 2005 Multipotential differentiation of adipose tissue-derived stem cells *Keio J. Med.* **54** 132–41

[23] Erisken C, Kalyon D M, Wang H, Ornek-Ballanco C and Xu J 2011 Osteochondral tissue formation through adipose-derived stromal cell differentiation on biomimetic polycapro-lactone nanofibrous scaffolds with graded insulin and beta-glycerophosphate concentrations *Tissue Eng.* A **17** 1239–52

[24] Panseri S, Russo A, Cunha C, Bondi A, Di Martino A, Patella S and Kon E 2012 Osteochondral tissue engineering approaches for articular cartilage and subchondral bone regeneration *Knee Surgery, Sports Traumatology, Arthroscopy* **20** 1182–91

[25] Benders K E, van Weeren P R, Badylak S F, Saris D B, Dhert W J and Malda J 2013 Extracellular matrix scaffolds for cartilage and bone regeneration *Trends Biotechnol.* **31** 169–76

[26] Xu T, Zhao W, Zhu J M, Albanna M Z, Yoo J J and Atala A 2013 Complex heterogeneous tissue constructs containing multiple cell types prepared by inkjet printing technology *Biomaterials* **34** 130–9

[27] Barron J A, Wu P, Ladouceur H D and Ringeisen B R 2004 Biological laser printing: a novel technique for creating heterogeneous three-dimensional cell patterns *Biomedical Microdevices* **6** 139–47

[28] Shim J H, Lee J S, Kim J Y and Cho D W 2012 Bioprinting of a mechanically enhanced three-dimensional dual cell-laden construct for osteochondral tissue engineering using a multi-head tissue/organ building system *J. Micromech. Microeng.* **22** 085014

[29] Lee J S, Cha H D, Shim J H, Jung J W, Kim J Y and Cho D W 2012 Effect of pore architecture and stacking direction on mechanical properties of solid freeform fabrication-based scaffold for bone tissue engineering *J. Biomed. Mater. Res.* A **100** 1846–53

[30] Melchels F P, Barradas A M, Van Blitterswijk C A, De Boer J, Feijen J and Grijpma D W 2010 Effects of the architecture of tissue engineering scaffolds on cell seeding and culturing *Acta Biomaterialia* **6** 4208–17

[31] Cao T, Ho K H and Teoh S H 2003 Scaffold design and *in vitro* study of osteochondral coculture in a three-dimensional porous polycaprolactone scaffold fabricated by fused deposition modeling *Tissue Eng.* **4** 103–12

[32] Seol Y J, Kang T Y and Cho D W 2012 Solid freeform fabrication technology applied to tissue engineering with various biomaterials *Soft Matter* **8** 1730–5

[33] Fedorovich N E, Schuurman W, Wijnberg H M, Prins H J, van Weeren P R, Malda J, Alblas J and Dhert W J 2011 Biofabrication of osteochondral tissue equivalents by printing topologically defined, cell-laden hydrogel scaffolds *Tissue Eng.* C **18** 33–44

Organ Printing

Dong-Woo Cho, Jung-Seob Lee, Falguni Pati, Jin Woo Jung, Jinah Jang and Jeong Hun Park

Chapter 12

Tissue engineering: soft tissue (the heart—cardiac muscle, cardiovascular tissue and heart valves)

12.1 Ischemic cardiac diseases

Cardiovascular disease is a significant cause of death worldwide, and it includes many problems that are related to atherosclerotic processes. Atherosclerosis develops when plaque builds up in the walls of arteries. It narrows the arteries, making blood circulation more difficult, causes blood clots to form, and ultimately can stop the blood flow [1]. This is a major cause of ischemic heart diseases. The strategies available to address end-stage ischemic heart failure are heart transplantation or the use of left ventricular assist devices (LVADs). However, heart transplantation is limited by a severe donor shortage (available to <0.1% of heart failure patients) and immune system rejection after transplantation. LVADs are available to supplement cardiac function, but for a limited amount of time [2].

A promising method to achieve functional improvement following myocardial infarction (MI) is to transplant cells into the injured area to replace the damaged tissue. The most established cell-based strategy is to deliver cells directly into the injured myocardium; however, 50–90% of injected cells are lost in dispensing, and over 90% of remaining cells die within one week of injection due to the hypoxic environment [2]. The poor survival rate is attributable to continued ischemia within the graft and surrounding tissues. In this regard, a tissue engineering approach could be an alternative, and this could offer additional physicochemical support directly to the injured heart tissue or to the implanted cells to aid functional repair.

Numerous approaches have been explored to facilitate cell retention, survival and integration into the host tissue within the damaged heart region. Recently, there has been much research into patch-type 3D cardiac tissue constructs, such as engineering heart tissue (EHT) [3], sponge-like macro-porous structures [4] and cell sheet-based scaffold-free structures [5]. These structures can attenuate left ventricular dysfunction by mechanically supporting and restoring the injured myocardium through the delivery of a large number of cells, biomolecules and other biological materials [6]. Indirect paracrine mechanisms may be evolved in the neovascularization and re-muscularization through the mobilization and activation of endogenous progenitor cells. The effects of engineered 3D cardiac tissue constructs have been clearly demonstrated in animal models and the clinical feasibility of patching such cardiac tissue constructs is supported by the successful transplantation of collagen sponges and skeletal myoblast-based cell sheets at the injured myocardium [7]. The benefits associated with engineered cardiac tissue construct application include: (1) the reduction of the infarct size through the provision of structural support; (2) the delivery of cells, biomolecules and cytokines secreted from the embedded cells to reduce apoptosis and promote angiogenesis; and (3) the stimulation of endogenous mechanisms (e.g., the epithelial-to-mesenchymal transition of resident or implanted progenitor cells), which can promote cell migration from the implanted patches towards the injured myocardium.

3D patch-type structures can be fabricated using various biomaterials, such as naturally derived materials (e.g., collagen, fibrin, and alginate) [8–10], dECM [11] and synthetic polymers [12, 13]. Heart tissue decellularized ECM (hdECM) has shown great promise for the replication of the intact cardiac tissue microenvironment and the repair of the heart after MI. Recently, Ott *et al* demonstrated the use of decellularized whole heart tissue as a functional scaffold and as a biomaterial in the field of cardiac tissue engineering [14]. Godier-Furnemont *et al* evaluated the efficacy of hdECM sheet, which was generated by slicing the decellularized heart tissue, with a combination of growth factors as a functional cell delivery platform [15]. Additionally, a pericardium-derived scaffold was developed by Rajabi-Zeleti *et al* using chemical crosslinking of a solubilized decellularized pericardium matrix (dPM). They demonstrated enhanced proliferation, viability, migration and differentiation using dPM [16].

The major advantage of a 3D printing-based scaffold is the creation of a porous architecture throughout the inner structure so that oxygen and nutrients can be supplied. Although this technology has been applied successfully in various tissue-engineering applications, generating large-volume tissues with high oxygen-consumption rates (e.g., cardiac, pancreas and liver tissue) remains a challenge. For example, an avascular tissue graft can accommodate approximately 2.8×10^5 cells mL^{-1} with no hypoxic phenomena after transplantation; however, the physiological cardiac tissue has a 70-fold higher cardiomyocyte density [17]. Gaetani *et al* introduced 3D cell printing of human cardiomyocyte progenitor cells (hCMPCs) with alginate bioink and printed lattice-shaped structures to provide pores (figures 12.1(a) and (b)) [18]. Although more consideration of the effects of different ECM components and growth factor modifications is still required, it could represent a first step for the development of an *in vitro*-created 3D printed cardiac construct. As a follow-up study, Gaetani *et al* used a mixture of hyaluronic acid and gelatin (HA/Gel) hydrogel as a bioink to evaluate the therapeutic potential of a 3D-printed cardiac patch composed of hCMPCs. The printed construct provided enhanced cell attachment and proliferation, and

(a)

(b)

(c)

(d)

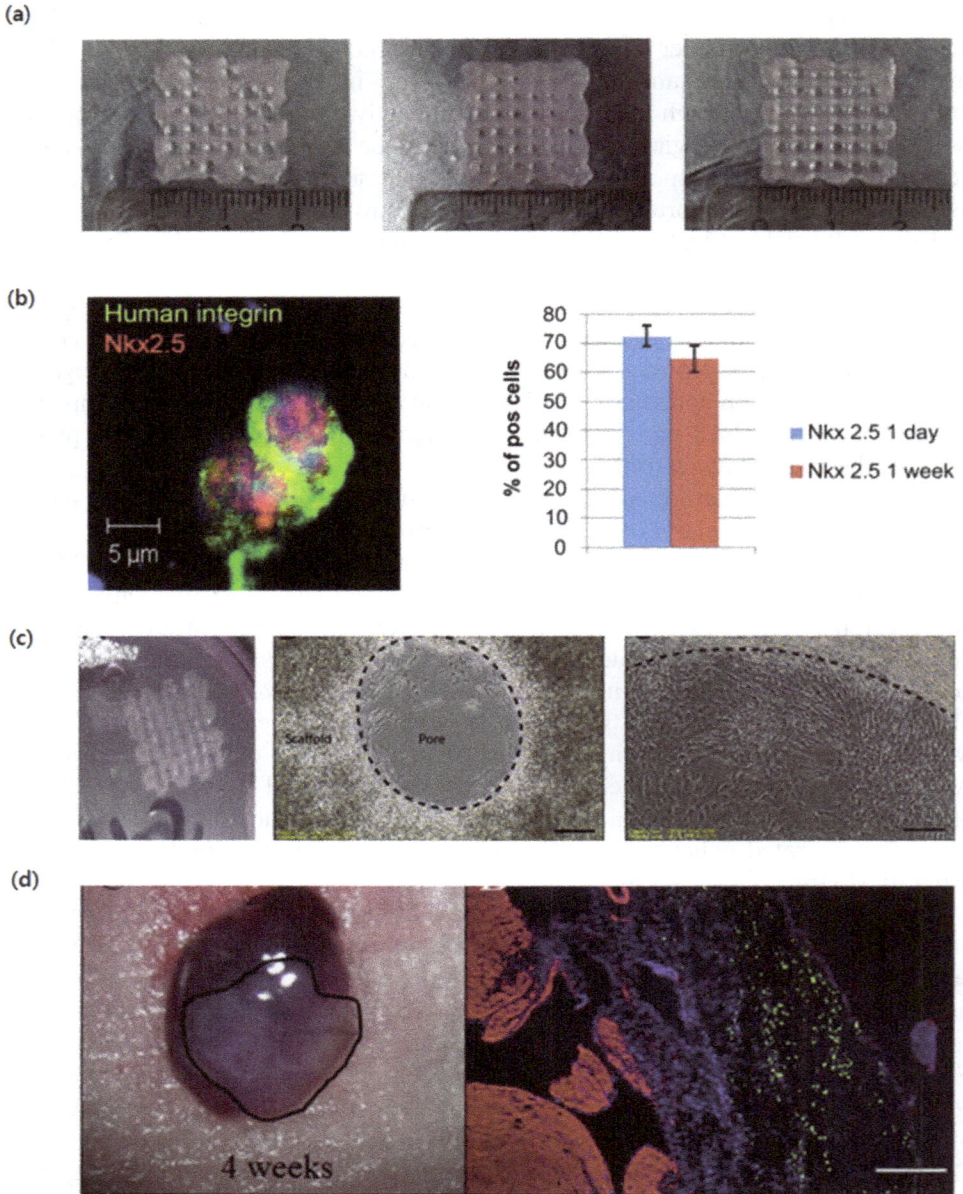

Figure 12.1. (a) Printed hCMPCs in 5% (left), 7.5% (middle) and 10% alginate scaffold (right). (b) immunofluorescence analysis of Nkx2.5 expression of printed hCMPCs after one week in culture (left) and quantification of Nkx2.5$^+$ cells (right). (c) 3D tissue printing of hCMPCs (left). Homogenously presented CMPCs in the structure one day after encapsulation; scale bar 200 μm (middle). Migrated CMPCs toward the outside of the printed patch after one week in culture; scale bar 100 μm (right). (d) Immunostaining of *in vivo* grafted patch four weeks after transplantation. (Red: cardiac troponin I, green: human lamin A/C and blue: DAPI). Reproduced with permission from [18, 19].

the cells in the printed construct retained cardiac differentiation capabilities *in vitro* for up to one month (figures 12.1(c) and (d)).

After implanting the construct, the left ventricle remodeling was attenuated and cardiac function was improved significantly [19]. This study highlighted the beneficial effects of the printing method for the effective delivery of cells directly to the injured site.

Ensuring sufficient vascularization of the engineered construct is essential for its long-term survival and a successful outcome. There are two approaches for printing vasculature in engineered tissues: (1) indirect printing using sacrificial materials that are removed by flow of the aqueous solution, leaving a vascular network behind; and (2) direct printing of a vascular network using endothelial cell sources. Recent studies have demonstrated the use of a sacrificial bioink to create vascular channels. Wu *et al* demonstrated the omnidirectional printing of 3D biomimetic microvascular networks by direct-write assembly of a sacrificial pluronic F127 bioink within a photocurable gel reservoir (figure 12.2(a)). This result shows the printability of a desired vascular network. After removing the sacrificial ink from the vasculature, the fluid can flow through the fabricated vasculature inside a large-volume construct [20]. Miller *et al* printed a 3D filament network composed of carbohydrate material-based water soluble bioglass for the rapid casting of patterned vascular networks in 3D large-volume tissues (figures 12.2(b) and (c)). They coated the fabricated vascular network with solubilized biocompatible polymer and then cast it into a 3D cell-laden

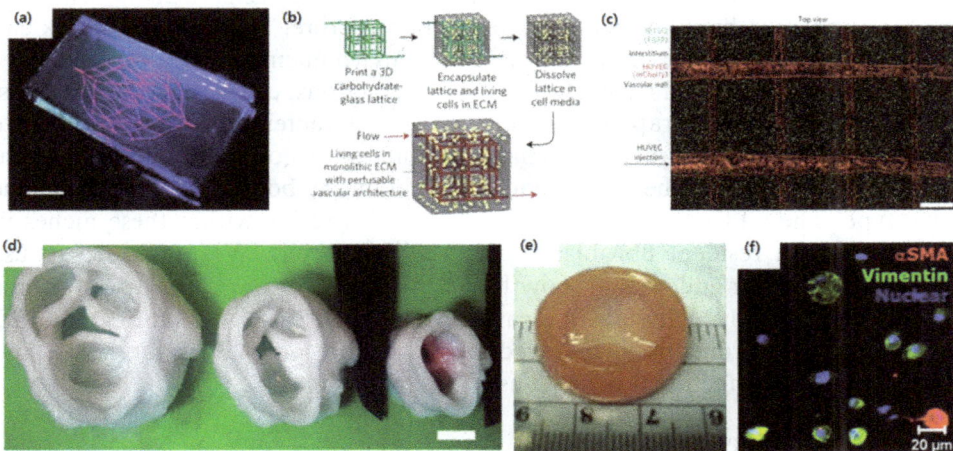

Figure 12.2. (a) Fluorescent image of a 3D microvascular network fabricated via omnidirectional printing of a fugitive ink (dyed red) within a photopolymerized Pluronic F127-diacrylate matrix (scale bar = 10 mm). (b) Schematic overview. An open, interconnected, self-supporting carbohydrate–glass lattice is printed to serve as the sacrificial element for the casting of the 3D vascular architecture. (c) Control of the interstitial zone and the lining endothelium of the vascularized tissue constructs (scale bar = 1 mm). (d) Scaffolds were printed with 700 MW PEG-DA at a different scale for fidelity analysis, where the inner diameters (ID) were 22, 17 and 12 mm. (e) Bioprinting of heart valve conduit with encapsulation of HAVIC within the leaflets. (f) Representative image of immunohistochemical staining for αSMA (green) and vimentin (red), and Draq5 counterstaining for cell nuclei (blue). Reproduced with permission from [14, 20–22].

natural ECM hydrogel. After washing the bioglass and flowing media through the hollow channel, embedded cells in the 3D ECM hydrogels showed higher cell viability and functionality [21]. The idea of directly printing blood vessels was confirmed by printing cell spheroids as building blocks for the fabrication of tubular-shaped structures [23]. However, direct vascular network printing is still under development as a concept. Nevertheless, microfabricated patterns (via soft lithography) demonstrated the usefulness of direct endothelial cell patterning to make a rapid and mature vascular network inside 3D large-volume tissue [24].

Several new findings, such as co-culture with endothelial cells or fibroblasts, delivery of angiogenic growth factors and extrinsic/intrinsic vascularization, have demonstrated the possibility of using capillary formation in 3D cardiac tissue constructs. With these successful developments, capillaries are relatively easy to grow inside an engineered tissue construct; but those capillary venules usually conduct blood very slowly. Thus, a next-step challenge will be the engineering of approaches to achieve hierarchical arterial and arteriolar formation in a high-metabolism 3D cardiac tissue construct populated with many cells. Additionally, rapid integration into the host myocardium is required to stop the LV remodeling process and to modulate the host environment via the influence of the 3D cardiac tissue construct. In this regard, Jang *et al* developed a 3D vascularized cardiac tissue construct using 3D cell-printing technology. It is a very versatile and simple pre-vascularization method, directly printing a vascular network within a cardiac tissue construct. They also investigated the improved cardiac differentiation of human c-kit$^+$ cardiac progenitor cells (hCPCs) when the developed hdECM bioink was used to print the structure. The hdECM bioink can provide a cardiac tissue homologous microenvironment, which can improve dynamic cellular behaviors, such as cell–cell interactions, cell–matrix interactions, soluble factors and topographical and mechanical characteristics, through intricate reciprocal molecular interactions. That environment is commonly called a 'stem cell niche'. Stem cell niches are found throughout the body, but depend on the tissue type. The CPCs in human cardiac tissue also reside within these niches in association with resident fibroblasts and amplifying cells. We attempted to use MSCs with vascular endothelial growth factor (VEGF) as the vascular cell source to generate a vascular network in a cardiac tissue construct. We fabricated a 3D pre-vascularized cardiac tissue construct by micropatterning CPCs and MSCs and demonstrated the functional benefits using animal tests.

12.2 Heart valve

Heart valve problems also require advanced treatment methods. In particular, the aortic valve usually regulates unidirectional flow of oxygenated blood to the myocardium and arterial system. The natural anatomical geometry and micro-structural complexity allows biomechanically and hemodynamically efficient function. The unique cell phenotypes in the compliant cusps contribute to the remodeling of valves for long-term durability within a demanding mechanical environment. The characteristics of atherosclerosis and the pathobiology of aortic valve stenosis are similar, but the final calcification is distinct. There are significant clinical needs; the mechanisms of normal valve mechanobiology and the mechanisms of the diseases should be studied. Tissue-engineered heart valves have increasing potential for use in valve disease therapy, but many hurdles in relation to current designs, cell sources and materials must be overcome through multi-disciplinary efforts [25].

Hockaday *et al* fabricated an aortic valve exhibiting complex 3D anatomy and heterogeneity using polyethylene glycol-diacrylate (PEG-DA) bioinks (700 or 8000 MW), supplemented with alginate bioink (figure 12.2(d)). High geometric accuracy was observed and the seeded porcine aortic valve interstitial cells survived for up to 21 days. The printed heart valve achieved an elastic modulus from 5.3 to 74.6 kPa and the larger printed valves had greater shape fidelity [13]. Duan *et al* used more biocompatible bioink (methacrylated hyaluronic acid (Me–HA) and methacrylated gelatin (Me–Gel)) than the previous study for making a cell-laden 3D printed heart valve structure (figures 12.2(e) and (f)). Printing accuracy was dependent on the ratio between the Me–Gel and Me–HA, and the trileaflet valve shape was replicated successfully using an optimized combination of the two materials [22].

12.3 Coronary arteries

Coronary artery disease continues to increase, and it already affects more than 16 million US adults. The symptoms may go unnoticed until the patient has a heart attack. The symptoms are managed through lifestyle modifications, drug therapy, percutaneous coronary intervention or coronary artery bypass grafting surgery (CABG). In a typical CABG surgery, internal thoracic arteries, radial arteries or saphenous veins are harvested from the rest of the body and then connected to the distal coronary arteries to provide blood flow. However, there are several limitations associated with CABG: sometimes accidental graft damage occurs during harvesting, it has relatively poor long-term patency and there are problems with donor site morbidity after surgery. Consequently, artificial coronary bypass grafts, such as polytetrafluoroethylene (ePTFE; Gore-Tex) [26] and woven polyethylene terephthalate (PET; Dacron), were developed to satisfy the clinical need. These materials provide an alternative to using isolated vasculature from the body. However, there are some important limitations to the use of these grafts: (1) improper mechanical properties and (2) low biochemical affinities.

Vascular tissue engineering has moved to a new level, with the ability to make artificial coronary bypass grafts, generating living blood vessel structures through tissue-engineering approaches. The tissue-engineered vascular graft should be non-thrombogenic and endothelialized, and should provide biomechanical properties similar to those of the native blood vessel. Norotte *et al* created scaffold-free engineered tissue constructs using 3D bioprinting technology. They printed cell aggregates at a desired location and then the cells naturally self-assembled in cylindrical shapes. This unique aspect of the method provided a new way to engineer vessels of distinct shapes and hierarchical trees with high controllability [23]. Li *et al* developed a vascular-like network using a double-nozzle assembling technique.

They printed adipose tissue-derived stromal cells (ASCs) encapsulated in a gelatin/ alginate/fibrinogen bioink to form a network construct in the 3D structure. This helped to create a vascularized structure in the hepatocytes, combined with a gelatin/ alginate/chitosan structure. The two parts were assembled and the ASCs were induced to differentiate into endothelial-like cells with endothelial growth factors [27]. In addition, Zhang *et al* introduced a new approach for fabricating vessel-like microfluidic channels that can be embedded in thick, engineered tissue or organ. They used a coaxial needle to dispense hollow hydrogel filaments and applied this strategy to engineer large-volume cartilage tissues [28].

References

[1] Lozano R *et al* 2013 Global and regional mortality from 235 causes of death for 20 age groups in 1990 and 2010: a systematic analysis for the Global Burden of Disease Study 2010 *The Lancet* **380** 2095–128

[2] Go A S *et al* 2014 Heart disease and stroke statistics–2014 update: a report from the American Heart Association *Circulation* **129** e28

[3] Zimmermann W-H *et al* 2006 Engineered heart tissue grafts improve systolic and diastolic function in infarcted rat hearts *Nat. Med.* **12** 452–8

[4] Madden L R *et al* 2010 Proangiogenic scaffolds as functional templates for cardiac tissue engineering *Proc. Nat. Acad. Sci.* **107** 15211–6

[5] Matsuura K, Utoh R, Nagase K and Okano T 2014 Cell sheet approach for tissue engineering and regenerative medicine *J. Controll. Release* **190** 228–39

[6] Ye L, Zimmermann W-H, Garry D J and Zhang J 2013 Patching the heart cardiac repair from within and outside *Circulation Res.* **113** 922–32

[7] Sawa Y *et al* 2012 Tissue engineered myoblast sheets improved cardiac function sufficiently to discontinue LVAS in a patient with DCM: report of a case *Surgery Today* **42** 181–4

[8] Shim J-H, Kim J Y, Park M, Park J and Cho D-W 2011 Development of a hybrid scaffold with synthetic biomaterials and hydrogel using solid freeform fabrication technology *Biofab.* **3** 034102

[9] Park J Y *et al* 2015 3D printing technology to control BMP-2 and VEGF delivery spatially and temporally to promote large-volume bone regeneration *J. Mater. Chem.* B **3** 5415–25

[10] Matsumoto T, Sasaki J-I, Alsberg E, Egusa H, Yatani H and Sohmura T 2007 Three-dimensional cell and tissue patterning in a strained fibrin gel system *PloS one* **2**:e1211

[11] Pati F *et al* 2014 Printing three-dimensional tissue analogues with decellularized extracellular matrix bioink *Nature Commun.* **5** 3935

[12] Kolesky D B, Truby R L, Gladman A, Busbee T A, Homan K A and Lewis J A 2014 3D bioprinting of vascularized, heterogeneous cell-laden tissue constructs *Adv. Mater.* **26** 3124–30

[13] Hockaday L *et al* 2012 Rapid 3D printing of anatomically accurate and mechanically heterogeneous aortic valve hydrogel scaffolds *Biofab.* **4** 035005

[14] Ott H C *et al* 2008 Perfusion-decellularized matrix: using nature's platform to engineer a bioartificial heart *Nat. Med.* **14** 213–21

[15] Godier-Furnémont A F *et al* 2011 Composite scaffold provides a cell delivery platform for cardiovascular repair *Proc. Nat. Acad. Sci* **108** 7974–9

[16] Rajabi-Zeleti S *et al* 2014 The behavior of cardiac progenitor cells on macroporous pericardium-derived scaffolds *Biomaterials* **35** 970–82

[17] Adler C and Friedburg H 1986 Myocardial DNA content, ploidy level and cell number in geriatric hearts: post-mortem examinations of human myocardium in old age *J. Mol. Cell. Cardiol.* **18** 39–53

[18] Gaetani R *et al* 2012 Cardiac tissue engineering using tissue printing technology and human cardiac progenitor cells *Biomaterials* **33** 1782–90

[19] Gaetani R *et al* 2015 Epicardial application of cardiac progenitor cells in a 3D-printed gelatin/hyaluronic acid patch preserves cardiac function after myocardial infarction *Biomaterials* **61** 339–48

[20] Wu W, DeConinck A and Lewis J A 2011 Omnidirectional printing of 3D microvascular networks *Adv. Mater.* **23** H178-H83

[21] Miller J S *et al* 2012 Rapid casting of patterned vascular networks for perfusable engineered three-dimensional tissues *Nat. Mater.* **11** 768–74

[22] Duan B, Kapetanovic E, Hockaday L A and Butcher J T 2014 Three-dimensional printed trileaflet valve conduits using biological hydrogels and human valve interstitial cells *Acta Bioa.* **10** 1836–46

[23] Norotte C, Marga F S, Niklason L E and Forgacs G 2009 Scaffold-free vascular tissue engineering using bioprinting *Biomaterials* **30** 5910–7

[24] Nichol J W, Koshy S T, Bae H, Hwang C M, Yamanlar S and Khademhosseini A 2010 Cell-laden microengineered gelatin methacrylate hydrogels *Biomaterials* **31** 5536–44

[25] Butcher J T, Mahler G J and Hockaday L A 2011 Aortic valve disease and treatment: the need for naturally engineered solutions *Adv. Drug Delivery Rev.* **63** 242–68

[26] Tu R H and Wang E 1989 Blood vessel replacements *Google Patents*

[27] Li S, Xiong Z, Wang X, Yan Y, Liu H and Zhang R 2009 Direct fabrication of a hybrid cell/hydrogel construct by a double-nozzle assembling technology *J. Bioactive Compatible Polymers* **24** 249–65

[28] Zhang Y, Yu Y and Ozbolat I T 2013 Direct bioprinting of vessel-like tubular microfluidic channels *J. Nano. Engg Med* **4** 020902

Dong-Woo Cho, Jung-Seob Lee, Falguni Pati, Jin Woo Jung, Jinah Jang and Jeong Hun Park

Chapter 13

In vitro tissue/organ models

In vitro 3D tissue models often provide better results in terms of drug/chemical screening than traditional 2D cell culture and animal models [1, 2], because 3D tissue models represent the spatial and chemical complexity of living tissues better than their 2D counterparts. These models enhance our understanding of tissue morphogenesis and also facilitate the development and screening of new therapeutics. Several types of 3D culture model have been developed using hydrogels of natural or synthetic origin that induce cells to polarize and interact with neighboring cells [3]. Such models possess different forms, including cells randomly interspersed in a matrix or clustered in self-assembling cellular microstructures, known as organoids [4]. However, many such models lack multiscale architecture and tissue–tissue interfaces (such as the interface between the vascular endothelium and surrounding connective tissue and parenchymal cells), which are essential for the function of nearly all tissues and organs [5].

Advances in 3D printing have enabled the direct assembly of cells and matrix materials to form structures that mimic native tissues, facilitating the development of *in vitro* tissue models for 3D biology, the study of disease pathogenesis and drug discovery [3]. Biological constructs of mm to cm size with intricate microstructures that include several cell types, biomolecules and biomaterials can be generated by organ printing [6]. Thus, *in vitro* 3D tissue models of human cells that mimic specific *in vivo* behaviors can be useful for the accurate prediction of drug/chemical responses and they reduce costs. Organ printing can create 3D constructs where cells are allowed to interact with adjoining cells; thus, their reactions to drug compounds are similar to those in the body. Cell printing techniques use the deposition of single cells or cell aggregates, where cells are generally loaded into gels for printing. The main advantage of this method is that the gels can be combined easily with additional components, such as drugs [7], biomolecules [8] or printed microfibers [9], to promote the formation of specific tissues as well as to improve the mechanical properties of the bioprinted constructs.

The bioprinted tissue possesses many features of the native tissue, including tissue-like cellular density, the presence of multiple cell types and key architectural and functional aspects [7]. The feasibility of constructing a physiologically relevant pharmacokinetic model using the 3D bioprinting process has been demonstrated [10]. In this approach, 3D bioprinted cell-encapsulated hydrogel-based tissue constructs were integrated directly onto a microfluidic device for continuous perfusion drug flow. Characterization of the bioprinted 3D tissue constructs showed predictable cell viability/proliferation results and enhanced functionality over traditional culture methods. San Diego-based Organovo Holdings, a biotech firm that designs and prints functional 3D human tissues with an inkjet-based bioprinter, is engaged in developing 3D bioprinted tissues for pharmaceutical and toxicological screening. Recently, Organovo developed its first 3D liver tissue for drug testing, making a breakthrough towards the commercial launch of a 3D liver tissue model [11].

Furthermore, *in vitro* disease models can be developed by fabricating constructs of diseased or dysfunctional human cells and these can be used in the testing of new drugs. Clinical trials are the most effective way to evaluate antineoplastic agents, although this method is not commonly used because of ethical and safety concerns. Thus, preclinical tumor models that mimic the physiological environments of tumors are worthwhile for studying tumor prognosis and antineoplastic drug screening [12–14]. *In vitro* 3D tumor models based on human cancer cells could be beneficial to accurately reproduce the characteristics of human cancer tissues [12, 15]. Various techniques, such as multicellular spheroids, cell-seeding 3D scaffolds, hydrogel embedding, microfluidic chips and cell patterning, have been explored for the production of 3D *in vitro* tumor models [12]. However, the techniques above are not adequate for simulating the complex 3D physiological tumor microenvironment and developing fully realistic 3D tumor models. Despite this, organ printing can be effective in producing complex structures that mimic physiological microenvironments for the development of *in vitro* disease/tumor models to aid in the study of disease pathogenesis [16].

Although there are few reports that deal with 3D printing of tumor models, the following study clearly demonstrates the enormous potential of the technique. For the development of *in vitro* cervical tumor models, 3D bioprinting of HeLa cells in gelatin/alginate/fibrinogen hydrogels was performed [17]. When compared with conventional 2D planar culture models, HeLa cells showed a higher proliferation rate in the printed 3D environment and tended to form cellular spheroids, whereas monolayer cell sheets were formed in the 2D culture. HeLa cells in 3D printed models also showed higher MMP protein expression and chemoresistance than those in 2D culture [17]. The new biological characteristics of printed 3D tumor models *in vitro* combined with novel 3D cell printing technologies are likely to advance 3D cancer study considerably in the future.

References

[1] Griffith L G and Swartz M A 2006 Capturing complex 3D tissue physiology *in vitro Nat. Rev. Mol. Cell Biol.* **7** 211–24

[2] Rangarajan A, Hong S J, Gifford A and Weinberg R A 2004 Species- and cell type-specific requirements for cellular transformation *Cancer Cell* **6** 171–83

[3] Kimlin L, Kassis J and Virador V 2013 3D *in vitro* tissue models and their potential for drug screening *Exp. Opin. Drug Discov.* **8** 1455–66

[4] Baptista P M, Siddiqui M M, Lozier G, Rodriguez S R, Atala A and Soker S 2011 The use of whole organ decellularization for the generation of a vascularized liver organoid *Hepatology* **53** 604–17

[5] Bhatia S N and Ingber D E 2014 Microfluidic organs-on-chips *Nat. Biotech.* **32** 760–72

[6] Fedorovich N E, De Wijn J R, Verbout A J, Alblas J and Dhert W J 2008 Three-dimensional fiber deposition of cell-laden, viable, patterned constructs for bone tissue printing *Tissue Eng. A* **14** 127–33

[7] Kaigler D, Silva E and Mooney D 2012 Guided bone regeneration using injectable vascular endothelial growth factor delivery gel *J. Periodontol.* **84** 230–8

[8] Wüst S, Godla M E, Müller R and Hofmann S 2014 Tunable hydrogel composite with two-step processing in combination with innovative hardware upgrade for cell-based three-dimensional bioprinting *Acta Biomaterialia* **10** 630–40

[9] Visser J *et al* 2015 Reinforcement of hydrogels using three-dimensionally printed microfibres *Nat. Commun.* **2015** 6

[10] Chang R, Nam J and Sun W 2008 Direct cell writing of 3D microorgan for *in vitro* pharmacokinetic model *Tissue Eng. C* **14** 157–66

[11] Organovo 2014 3D *Human Liver Tissue Testing Services* (San Diego, CA: Organovo)

[12] Kim J B 2005 Three-dimensional tissue culture models in cancer biology *Semin. Cancer Biol.* **15** 365–77

[13] Jordan M A, Toso R J, Thrower D and Wilson L 1993 Mechanism of mitotic block and inhibition of cell proliferation by taxol at low concentrations *Proc. Natl Acad. Sci. USA* **90** 9552–6

[14] Ellingsen C, Natvig I, Gaustad J-V, Gulliksrud K, Egeland T M and Rofstad E 2009 Human cervical carcinoma xenograft models for studies of the physiological microenvironment of tumors *J. Cancer Res. Clin. Oncol.* **135** 1177–84

[15] Bissell M J and Radisky D 2001 Putting tumours in context *Nat. Rev. Cancer* **1** 46–54

[16] Mironov V, Trusk T, Kasyanov V, Little S, Swaja R and Markwald R 2009 Biofabrication: a 21st century manufacturing paradigm *Biofabrication* **1** 022001

[17] Zhao Y *et al* 2014 Three-dimensional printing of HeLa cells for cervical tumor model *in vitro Biofabrication* **6** 035001

Chapter 14

Future work

Organ printing is a vibrant research area in which a number of pioneering results have been obtained in the last decade. Although organ printing is currently at a nascent stage, this emerging technology has the potential to advance tissue engineering towards functional tissue and organ fabrication for clinical transplantation, ultimately addressing the shortage of organs and saving lives. However, there are some important aspects that should be addressed before that can happen. For example, current technology should be upgraded for standardized, scalable fabrication methods for the robotic delivery of cells and matrix materials. Researchers should consider the following critical requirements for improved use of the technique: (1) better designs for nozzles and cartridges; (2) novel bioprocessable and functional bioinks that support high cell density; (3) strategies for integrating branched vascular trees within bioprinted tissue for occlusion- or leak-free perfusion; (4) novel technologies for multi-bioink, multi-scale hybrid bioprinting processes; (5) enhanced tissue maturation technologies for faster tissue development; and (6) the design of bioreactors to facilitate postprocessing remodeling after printing.

To date, most published reports have been proof-of-concept studies that demonstrate the fabrication of simple tissues with basic structures. Only a few studies have investigated process parameters for either predictions or optimization strategies in a systematic way. Thus, studies targeted towards understanding the relationship between process parameters and the structure and functions of the printed constructs are required urgently. Moreover, modern fabrication schemes rely on mathematical modeling and computer simulations to optimize process design and predictions [1, 2]. Using computer simulations, the constructs can be predicted and thus optimized before printing. However, this approach needs more attention regarding the precise use of new biological processes and 3D tissues.

Organ printing can be useful for the development of tissue/organ models to evaluate the efficacy of pharmaceutical drugs or biomolecules. Although there has not been much research carried out in this direction, huge possibilities are emerging with the advent of a ban on animal-tested products (in particular, the one imposed by the European Commission) [3]. Digital bioprinting may provide another example of a case in which the biomaterial is labeled with magnetic nanoparticles and can be positioned in a controlled manner due to magnetic interactions [1]. It is predicted that in the near future the use of organ printers will be as common as the use of microscopes in biological, academic, clinical and industrial laboratories [4].

A comparatively new concept is that of 4D bioprinting, where the rapid fusion, folding and remodeling of cell aggregate-based printed pre-tissues takes place, leading to the fabrication of living tissues during a shorter period of *in vitro* culture. This is particularly promising as it allows living tissues to be generated in shorter times, enabling bioprinting in the fourth dimension, time.

14.1 Conclusions

In recent years, organ printing has been much investigated for the generation of 3D cellular constructs using complex designs and novel bioinks. This can be seen from the number of publications and the creation of new societies. Although these techniques are still in their infancy, they have the potential to overcome the problems associated with the current scaffold-based tissue regeneration process. There are basically three kinds of bioprinters available for organ printing; however, bioprinters capable of printing microstructures with nano-scale features and integrated microvasculatures would be worthwhile for the realization of complex tissue models. Organ printing relies on printing the functional tissue construct by mimicking the physical or chemical attributes of the native ECM or through designing and developing suitable bioinks. It is a sophisticated technique for the production of constructs with desired geometries, mechanical properties and functionalities, using different materials, and has the potential to effect the controlled placement of viable cells. The functional restoration of living tissues and organs can be further enhanced by the spatiotemporal delivery of bioactive factors or signaling molecules, and future studies should consider integrating these factors within bioprinted constructs. Nevertheless, organ printing will open up new avenues to scalable and reproducible mass production of tissue/organ precursors and, ultimately, solve the tissue/organ shortage.

References

[1] Mironov V, Trusk T, Kasyanov V, Little S, Swaja R and Markwald R 2009 Biofabrication: a 21st century manufacturing paradigm *Biofabrication* **1** 022001
[2] Fabien G, Vladimir M and Makoto N 2010 Bioprinting is coming of age: report from the International Conference on Bioprinting and Biofabrication in Bordeaux (3B'09) *Biofabrication* **2** 010201
[3] European Commission 2015 *Ban on Animal Testing* http://ec.europa.eu/growth/sectors/cosmetics/animal-testing/index_en.htm
[4] Mironov V 2003 Printing technology to produce living tissue *Exp. Opin. Biol. Ther.* **3** 701–4